Profit Improvement through Supplier Enhancement

Systems Innovation Series

Series Editor

Adedeji B. Badiru

Air Force Institute of Technology (AFIT) – Dayton, Ohio

PUBLISHED TITLES

Additive Manufacturing Handbook: Product Development for the Defense Industry,
Adedeji B. Badiru, Vhance V. Valencia, & David Liu

Carbon Footprint Analysis: Concepts, Methods, Implementation, and Case Studies,
Matthew John Franchetti & Defne Apul

Cellular Manufacturing: Mitigating Risk and Uncertainty, *John X. Wang*

Communication for Continuous Improvement Projects, *Tina Agustiady*

Computational Economic Analysis for Engineering and Industry, *Adedeji B. Badiru &
Olufemi A. Omitaomu*

Conveyors: Applications, Selection, and Integration, *Patrick M. McGuire*

Culture and Trust in Technology-Driven Organizations, *Frances Alston*

Design for Profitability: Guidelines to Cost Effectively Management the Development Process
of Complex Products, *Salah Ahmed Mohamed Elmoselhy*

Global Engineering: Design, Decision Making, and Communication, *Carlos Acosta, V. Jorge Leon,
Charles Conrad, & Cesar O. Malave*

Global Manufacturing Technology Transfer: Africa–USA Strategies, Adaptations, and Management,
Adedeji B. Badiru

Guide to Environment Safety and Health Management: Developing, Implementing, and
Maintaining a Continuous Improvement Program, *Frances Alston & Emily J. Millikin*

Handbook of Construction Management: Scope, Schedule, and Cost Control,
Abdul Razzak Rumane

Handbook of Emergency Response: A Human Factors and Systems Engineering Approach,
Adedeji B. Badiru & LeeAnn Racz

Handbook of Industrial Engineering Equations, Formulas, and Calculations, *Adedeji B. Badiru &
Olufemi A. Omitaomu*

Handbook of Industrial and Systems Engineering, Second Edition, *Adedeji B. Badiru*

Handbook of Military Industrial Engineering, *Adedeji B. Badiru & Marlin U. Thomas*

Industrial Control Systems: Mathematical and Statistical Models and Techniques,
Adedeji B. Badiru, Oye Ibidapo-Obe, & Babatunde J. Ayeni

Industrial Project Management: Concepts, Tools, and Techniques, *Adedeji B. Badiru,
Abidemi Badiru, & Adetokunboh Badiru*

Inventory Management: Non-Classical Views, *Mohamad Y. Jaber*

Kansei Engineering — 2-volume set
- Innovations of Kansei Engineering, *Mitsuo Nagamachi & Anitawati Mohd Lokman*
- Kansei/Affective Engineering, *Mitsuo Nagamachi*

Kansei Innovation: Practical Design Applications for Product and Service Development,
Mitsuo Nagamachi & Anitawati Mohd Lokman

Knowledge Discovery from Sensor Data, *Auroop R. Ganguly, João Gama, Olufemi A. Omitaomu,
Mohamed Medhat Gaber, & Ranga Raju Vatsavai*

Learning Curves: Theory, Models, and Applications, *Mohamad Y. Jaber*

Managing Projects as Investments: Earned Value to Business Value, *Stephen A. Devaux*

PUBLISHED TITLES

Profit Improvement through Supplier Enhancement

By
Ralph R. Pawlak

CRC Press
Taylor & Francis Group
Boca Raton London New York

CRC Press is an imprint of the
Taylor & Francis Group, an **informa** business

CRC Press
Taylor & Francis Group
6000 Broken Sound Parkway NW, Suite 300
Boca Raton, FL 33487-2742

© 2017 by Taylor & Francis Group, LLC
CRC Press is an imprint of Taylor & Francis Group, an Informa business

No claim to original U.S. Government works

Printed on acid-free paper

International Standard Book Number-13: 978-1-138-70243-1 (Hardback)

Library of Congress Cataloging-in-Publication Data

Names: Pawlak, Ralph R., author.
Title: Profit improvement through supplier enhancement / Ralph R. Pawlak.
Description: Boca Raton, FL : CRC Press, 2017. | Series: Industrial
innovation series
Identifiers: LCCN 2016047889| ISBN 9781138702431 (hardback : alk. paper) |
ISBN 9781315203614 (ebook)
Subjects: LCSH: Industrial procurement--Quality control. | Production control.
Classification: LCC HD39.5 .P38 2017 | DDC 658.7--dc23
LC record available at https://lccn.loc.gov/2016047889

Visit the Taylor & Francis Web site at
http://www.taylorandfrancis.com

and the CRC Press Web site at
http://www.crcpress.com

Dedication

To my sons Glen and Ralph:
You have become better fathers
than I could have ever been.

Contents

Preface

This book, titled *Profit Improvement through Supplier Enhancement*, has an intended readership composed of owners, managers, supervisors, quality engineers, controllers, consultants, trainers, purchasing agents, shipping agents, engineering students, and others involved in any company or endeavor that has manufacturing or service facilities, whether in-house or outsourced.

One of two previous books (*Industrial Problem Solving Simplified*, Apress, 2014) contained a plan to evaluate and solve specific manufacturing or service problems. It did not deal directly with supplier enhancement activity. That book focused on defining problems, characterizing faults and using concept sheets. It described the steps necessary to develop a plan of attack and the means to collect data and generate clues to enable solutions to industrial problems. The book also described the use of innovative tools and the necessity to establish consistent work and work reviews to improve problem-solving skills. A second book (*Solving Complex Industrial Problems without Statistics*, CRC Press, 2016) related real-life experiences of analyzing and solving supplier, service, and manufacturing problems. These experiences led to protocols for solving industrial problems without the use of statistical analysis.

These steps, however, are not all that is required to improve quality and to improve the profitability of a manufacturing or service organization. An important addition to a successful organization is the development of a supplier base that will help improve or maintain its cost-effectiveness and profitability. This customer–supplier relationship is not fully realized in most organizations. In a majority of cases suppliers are treated as a necessary evil rather than as a business partner. Unfortunately, a lack of appreciation of this realization prevents many organizations from reaching their full potential in quality and profitability. An enhanced supplier base can contribute to a smooth symbiotic relationship that enhances both the supplier's and the customer's performance. This can allow both to achieve improved economic and performance levels.

Most customer–supplier relationships can be effective in reducing the related costs of both parties. Ultimately, the benefits realized from the use of supplier enhancement does affect the bottom lines of both the supplier and the customer. The information accessible herein is presented in its most simple form. It is geared to practitioners with what I hope are simple, clear, and concise case studies and applicable tools.

But before the material is quantified and presented, let me address what the book does not contain. This book does not provide any complex multicriteria optimization methods or models. It does not contain any significant research findings that can be utilized to elevate a decision-making policy in a supply chain. It does not provide any risk modifications or qualification models to improve or involve supplier process improvement strategies. It does not contain any complex enhancement models or production planning sequences or methods to improve supplier material availability.

Rather, it contains information that was developed working with a supplier base of over 200 individual suppliers in various industries in the United States, Canada, Europe, Philippines, Mexico, and China. All of the descriptions and samples were derived from actual supplier performance improvement steps established to eliminate supplier problems. These conditions are described in their essence and are presented in worksheets, audit items, or line items. Each chapter contains a summary, and the final chapter contains an overall summary of the materials presented. This allows those not familiar with the simplified methods to apply them to their circumstances even if they have no prior problem-solving experience. It includes work lists, diagrams, plan descriptions and worksheets that contain recommended considerations for suppliers to use to make the required improvements in their problem-solving attempts.

This book then deals with the improvement of suppliers, which is an unrecognized area of low-hanging fruit to increase a company's top and bottom lines. The enhancement of suppliers can be accomplished in a series of steps when conditions warrant intervention. The enhancements can also be generated through direct quality mentoring when the supplier does not possess the basic skills or abilities to adequately address impending problems. Consequently, this book has been penned to provide information and guideline activities that will aid any supervisor, owner, leader, or executive to achieve and attain improved profitability and competitiveness in the world market.

Working with sundry quality and purchasing managers over the years, it was found that a philosophy of treating suppliers as business partners rather than as adversaries brought about the best results. A productive relationship between the customer and inexperienced suppliers can be achieved by mentoring those suppliers to practice the steps that will be discussed. A series of proven best practices, processes, systems, and tools were accumulated over a period of many years by resolving supplier problems in the automotive, chemical, electronic, gear, engine, casting, assembly, forming, clothing, juvenile products, and toy industries. These practices are discussed here with the intention of improving the supplier base.

It is intended that the presentation of these easily applicable steps will help facilitate improvements in your supplier performance. The improved performance of your supplier base can positively affect the top- and bottom-line profitability of any company, department, store, or service. If customers espouse and manage these basic concepts, the organization (large or small) will be more capable of improving profitability than undisciplined competitors.

Supplier enhancement for some may seem to be a dry subject. To make the book more exciting and explicit, it is loaded with actual supplier examples and problems to compliment the information presented and to make it more easily understood, interesting, and useful. The presentation of the numerous forms provided is intended to allow easy access for those wishing to establish systems that may be applicable to their organization. The reader may desire to write notes or thought starters in the margin to aid in any future system applications.

You are invited to join those managers that are resolved to providing improved profits for their organization by enhancing supplier performance and by applying the plans, examples, and worksheets provided within this book.

Acknowledgments

I wish to thank the following individuals for their aid and expertise that they provided me in the production of this book:

Cindy Carelli, the executive editor at CRC Press, who recognized the potential of the contents of the book and oversaw its progress.

Jay Margolis, the production editor at CRC Press, who coordinated the project through the necessary preparation steps.

Michelle van Kampen, the Project Manager at Deanta Global, who coordinated the editing and layout of the materials and prepared the final product for consumption.

Prabhu Venkatesan at Deanta Global for providing the proof pages in a timely manner.

Matthew Grundy Haigh, the copy editor, who enabled the presented materials to be clear and understandable.

Sandra Robinson, the graphics designer who illustrated the book cover to be representative of the enclosed material.

Nakash, the editorial assistant to Cindy Carelli, for her follow-up and assistance.

About the Author

Ralph R. Pawlak has experienced an eventful career in the quality, manufacturing, and managerial fields. He has earned degrees in industrial technology (AAS) from Erie Community College, an industrial engineering degree from the General Motors Institute (BIE), and a master of education degree from the University of Buffalo (EdM). He has applied some of these experiences as an adjunct professor at Erie Community College.

He gained extensive experience within the United States, Canada, Europe, Hong Kong, and China in various quality engineering and management activities. Most of these activities involved the interactions with suppliers that led to the information presented within this book. The most memorable of these activities involved solutions to dire problems that had a direct impact on production schedules and both top- and bottom-line profits.

His two previous books, *Industrial Problem Solving Simplified* (Apress, 2014) and *Solving Complex Problems without Statistics* (CRC Press, 2016), did not address supplier improvements and enhancements. Hence, this book was written in a desire to pass on what he has learned so that others may benefit from his experiences.

chapter one

Introduction

One of the key causes of losses in profitability is the lack of attention to supplier effects on the top and bottom line. This may be due to the relative inexperience of administrators or the lack of interest in what may appear to be a mundane subject. It is important to recognize that there is potential for profit improvement in capturing the low-hanging fruit obtainable through supplier improvement. Therefore, businesses large and small could improve profitability through the application of supplier enhancement modifications.

Perceptive managers recognize that commonplace problems can be initiated by both internal and external suppliers. There are conventional suppliers of services or products that are employed by other organizations, whereas internal suppliers are those that are encountered within a home organization. Both of these supplier types will be discussed in detail in the following chapters. The recommendations in this book are applicable to both internal and external supplier bases.

The primary cause of many project, manufacturing or service problems is the failure to affect corrective change once a problem has been recognized. This is true for problems created by both internal and external supplier sources as well. Unless providers are held accountable for the provision of meaningful controllable solutions to the problems encountered, you can be relatively confident that the same problems will reappear in the future. Each time any one of these repeat problems occurs someone ultimately incurs an unnecessary monetary loss.

This book provides tools that can be employed to gain control of problematic supplier situations. These tools are applicable to all types of supplier problems whether they are created at Tier 1, Tier 2 or a local interdepartmental location. The tools are useful in determining both the cause of supplier-related problems as well as the actions that are to be initiated to prevent the recurrence of similar problems.

Features

Some of the features of this book are as follows:

- Presents real-life examples from over 200 suppliers
- Provides steps for resolutions to supplier problems

- Shows corrective considerations applicable to internal or external supplier problems
- Explains the methods used to discover supplier problems
- Specifies the information required at supplier quality meetings
- Outlines corrective steps to be required of suppliers
- Defines the methods to be used at supplier symposiums
- Contains forms that identify clues to resolve problems
- Explains sample audits to attain and control process and product flow
- Describes the methods used to manage improvements
- Sets the path for optimal supplier problem resolution
- Proposes methods to help unproven suppliers
- Provides audit forms for supplier rating and evaluation
- Explains the Trilogy of Supplier Enhancement concepts of initial material checking (*Do Check*), ensuring quality (*Don't Create*), and establishing criteria for securing defective materials (*Do Hold*)
- Provides basic questions to identify the sources of problems and to give an indication of the necessary corrective actions to be taken
- Provides adequate spacing between paragraphs to allow the reader to add notes that may be personal or beneficial to their individual circumstances

Means of supplier improvement

The following sequence of information is provided to deliver the route by which the reader can understand and improve their performance through supplier enhancement. The main path of understanding is best described as follows:

- The differences between internal and external suppliers and the common methods to improve both performance and relationships are provided.
- The book explains how to use simple techniques to identify and simplify supplier problems and how to innovate significant solutions.
- Information is given to delineate the differences between qualified suppliers and those who are less experienced.
- The presented material supplies an understanding of how supplier improvements can substantially improve top- and bottom-line performance.
- Forms and examples illustrate how to use a quality management approach to facilitate supplier improvements.
- A series of worksheets are presented that can be applied to the solution of specific supplier problems.
- An appendix is provided to define terms with which the reader may not be familiar.

Allow me to repeat the following sentence, as it is intended for those who may have not read the Preface. Since supplier enhancement may seem to be a dry subject to some, the book is loaded with actual supplier examples, photographs, audits, lists, and problems to compliment the information presented and make it more easily understood, interesting, and useful. The book also contains a significant number of example audit worksheets that have been applied to different departments in various countries and locations.

Flow path of the book

The contents of the book are presented in the following sequence so that the reader can secure the information efficiently. Some terms used within the book may be new to the reader. Most will be self-explanatory, but novel words will be defined in the Appendix.

Chapter 1: Introduction. This chapter provides the basis for the need to address supplier enhancement. It also contains an outline or flow path of the information that is presented. The investigation involves looking at the conditions that exist during discussion and evaluation in order to act on supplier problems. It identifies some features that affect the customer–supplier relationship. These features are listed as background material. The study addresses the conditions that were found to exist and proposes the necessary corrective actions. It also contains an outline or flow path of the information that is presented.

Chapter 2: Problem history. This chapter introduces the need for generating supplier improvements. It contains a presentation of a sample of problems created by suppliers. The chapter relates the importance of collecting data to pinpoint problems and the methods of resolving those problems that will be explained in the ensuing chapters.

Chapter 3: Need for enhancement. This chapter relates the information gleaned during a study to address supplier problems and subsequent improvements. It contains what was found to be the best and worst actions practiced by the supplier base, the problems encountered and the key actions taken to soundly form and establish corrective practices.

Chapter 4: Supplier credentials. This chapter describes the types of suppliers that are encountered by a customer. It describes the differences and similarities of those that may be involved in the manufacture of a product or the provision of a service. Examples are given of supplier traits, with a focus on the differences that are to be considered. Descriptions are given of instances of supplier inefficiency, inaction or impertinence that result in ineffectiveness and dollar loss. The chapter contains examples of the destructive activities of unfit suppliers. It introduces some supplier control and improvement concepts that will be discussed in later chapters.

Chapter 5: Supplier elements. This chapter contains a description of the preferred supplier characteristics that should be considered when acquiring a supplier. "Mom and pop" shops versus small and large manufacturers are examined. The mammoth leader corporations are also studied in terms of their use of quality tools, their responsiveness, and their achievements. The severity of supplier-induced problems is described as a function of their expertise. Also discussed are the necessary steps to be taken to achieve supplier approvals and enhancements.

Chapter 6: Course of action. This chapter describes those actions that can be taken depending on the supplier's circumstances. The chapter suggests an innovative technique for accepting supplied components or assemblies without any additional in-house inspection at the receiving location. It also suggests specific employee actions to ensure product integrity. Included is an example of a problem resolution sheet that can be used to drive corrective actions. These actions are determined by conducting different types of meetings, which are described. The dialog indicates actions that can be taken to resolve any supplier problems that may arise. The chapter relates the importance of helping suppliers to utilize design failure mode and effect analysis (DFMEA), process failure mode and effect analysis (PFMEA), and control plans to improve their quality systems. Unqualified suppliers that do not currently have use of these tools must be developed to ensure successful operations.

Chapter 7: Record keeping. A chapter on record keeping is provided to focus attention on supplier errors and problem resolutions. It is important to define the problem, to describe the failure mode encountered, the process flow plan, the product containment method, and the date code interpretation. All of these are critical to creating and maintaining correct records. The chapter also lists three important supplier attributes that should be available at the supplier location. The use of DFMEA, PFMEA, and control plans will be instrumental in their improvement and development. Suppliers that do not possess these attributes will require guidance and help to develop.

Chapter 8: Supplier readiness. Process and supplier improvements are created by rectifying adverse conditions and addressing them in the product or process control plan. Supplier readiness is a means of determining the supplier's inclination to deliver the desired product on time, in the amounts required and in an acceptable condition. This chapter deals with those considerations that are to be evaluated when a supplier is being considered for a new activity or contract. Otherwise, the information can be used to qualify a new supplier for approval. In addition, the observations can be used to evaluate why a supplier may be supplying a product that is not satisfactory. This is a good start for those that have no formal system and can provide insight into areas that were not formally considered. (Forms and examples are included.)

Chapter 9: Opportunities for improvement. The center of attention in this chapter is how to provide some less experienced suppliers with the expertise to improve their systems. The problem discussed is that of receiving mixed parts, which can cause line jams, improper assembly, and related safety problems. The conditions explained will lead to improvements in both the supplier's and the customer's systems and the quality of the product that the customer receives.

Chapter 10: Problem resolution aids. This section provides a simple method to allow inexperienced or inept suppliers to improve their problem-solving activities. The chapter more fully explains why defective materials should not be accepted at the job site. Once acceptable materials have been received, it is necessary to perform specific operations with the methods and tools provided. After any operation is satisfactorily accomplished, the work should be viewed to ensure that it meets requirements and that it can be forwarded to the next customer or operator. In no event should defective material be forwarded to the next operation. It should be sequestered and placed on hold.

Chapter 11: System evaluation. This chapter is one of the most salient and important in the book for supplier improvement and enhancement. It provides an audit guide to be used at a supplier's location to establish and rate their process implementations. It can be used in association with the tools previously discussed and the DFMEA, PFMEA, and control plans that are instituted at the supplier's location. The forms provided can be copied and used as a reference or for conducting supplier evaluations internally or by outside sources. This information can also be used to differentiate the qualifications of different suppliers.

Chapter 12: Other requirements. It follows that negative results obtained during audit observations must be corrected. Some of the actions required to provide process corrections can be warranted by the presence of other systems or practices. These requirements are addressed in the chapter. It defines specific criteria to demand from suppliers when manufacturing or service problems arise. Topics include corrective actions, compliance with customer requirements, the documentation and implementation of the quality system, the availability of records and information, management participation and action, new contract requirements, workstation observation, quality process pointers, manufacturing metrics, readiness, operational issues, and finally, a plant visit and tour of the facilities.

Chapter 13: Tools for improvement. This chapter has information that will allow suppliers to improve their problem-solving abilities. This section is necessary because some suppliers lack the expertise or experience to solve complex problems. It befits the customer to aid them in their endeavors. This mentoring can be accomplished via actual meetings or conferencing with suppliers at the customer's or supplier's location when dire problems arise. These types of analyses might be applied as a requirement when problematic conditions appear. Some simple, useful tools that have been applied in various industries are presented for their use in preventing these recurrences.

Chapter 14: Useful audit criteria. This chapter is presented for the convenience of those who may now find it wise to establish or improve their supplier enhancement efforts. Examples of mentoring the supplier in problem resolution using audit observations are provided. The information is presented in categorical form to be used as a basis for the creation of pertinent criteria that must be considered for the respective industry being serviced. Because all industries are not identical, there are individual traits that should be developed by those that are involved. Since many suppliers lack the technical ability to analyze and identify the cause of a problem, this section provides multiple tools that can aid their efforts in problem identification or process improvement. This chapter provides some initial audit worksheet considerations that are relevant to most suppliers and which should be included in any proposed criteria development.

Chapter 15: Summary. The summary chapter reviews the important concepts that have been presented.

Appendix. The appendix defines terms used throughout the book.

Many illustrations, tables, graphs, photographs, illustrations, explanations, and worksheets are included to provide clarity and ease of use. The numerous usable audit forms provided can be applied to the customer base to improve profitability. The contents will increase your understanding of the supplier–customer relationship and will improve your performance in dealing with internal or external customer problems.

I did not have the opportunity to find most of the information presented in the book during my formal educational experience. I sincerely hope that it can be of benefit to you in your endeavors.

With that being said, let us now evaluate the customer–supplier relationship and the supplier enhancement considerations that can help to improve profits for both the customer and the supplier base. Initially, then, let us consider some examples of historical problems that have dogged suppliers and created customer consternation.

chapter two

Problem history

The previous chapter dealt with an introduction to the process of supplier enhancement. It also listed the features and the means of improving supplier performance for those that require it. A roadmap or flow path was presented to allow the reader to peruse the material for relevance and consequence to their needs.

This chapter describes the observations necessary for the undertaking of supplier enhancement. It has been written to transfer information about the methods that can be used to improve profitability through the application of supplier modification. Suppliers can and do affect the profitability of their customers. What I propose to transfer is the knowledge gained over many years of mentoring suppliers in various industries. Some organizations were serviced by over 200 different suppliers providing inputs ranging from raw materials, services, parts, or revolutionary systems. The reader should be able to easily assimilate the information presented so that it can be applied to their various circumstances.

The methods and systems that are contained within were found to be applicable in supplier situations in the United States, Canada, China, Hong Kong, Mexico, the Philippines, and Italy. These worldwide facilities manufactured automobile components, electronics, castings, gears, children's strollers, adolescent clothing, and juvenile toys, with either metal or plastic machined, formed, or assembled components. The numerous suppliers provided the services that were required to fulfill their contracts with the customer that was represented. This list illustrates that the processes described can be used anywhere to mentor suppliers that provide diverse products or services.

Supplier errors

Suppliers not only affect profitability, they can also create untold hardships and failures in their services. Consequently, they must be managed and controlled in order to ensure the safety of the product or service that they provide. Take, for instance, the following examples:

1. A bicycle supplier provided unsafe products to its wholesaler. This occurred because many of the interior suppliers, those within the organization, failed to be responsible. Unsafe conditions were

caused in the following sequence of events. The events followed the manufacturing process. In the first operation the front fender blank was create by punching a form from a roll of metal sheet stock, as shown in Figure 2.1.

However, the blanking operation forming the front wheel fender outline contained a sharp fin during the punching operation caused by a worn guide pin and bushings. (A fin is a sharp burr or metal extension that is not part of the desired product.) This piece was processed at the next operation to form the curvature of the fender, as shown in Figure 2.2.

This part was transferred to the painting department without being processed through a deburring operation to remove any unsafe sharp edges or burrs. After this the part was painted and assembled onto a bicycle frame. A final inspection did not catch the unsafe condition and it was shipped to the wholesaler. Once received, the bicycle was placed on display for sale at a reputable seller location.

Now, recall that this product was formed and processed internally by one manufacturer. There were no Tier 1 outside suppliers involved in its assembly. So let's relate why this example is relevant here as a supplier issue. Note that the individual operations within this company could also be considered to be Tier 1 and Tier 2 suppliers to the assembly department.

It is important that anyone providing a service check their operation and performance to ensure that the component, function, or service meets requirements. So each internal (in-house) supplier should ascertain and make certain that the function they fulfill is acceptable. *Don't create a defective product. Take no action or inaction that would cause a defect to exist.* (Don't Create.) Check the material input and output and apply corrective actions as they may be required. In this case the employee did not create a fin-free fender blank. The employee involved did not check the quality of the work accomplished and may have continued forming defective fender blanks.

Figure 2.1 Top and side view of a desirable fender blank.

(a)   (b)

Figure 2.2 Sharp burr on the periphery of a fender: (a) front view, (b) side view.

In addition, the painting, assembly, and inspection operations were all inattentive in providing their functions without adequately perusing the product that they received. The defect was passed through all of the stations without the defect being recognized or corrected. *Do perform a cursory check to examine for obvious defects on all workstation inputs.* (Do Check.) This includes external and internal suppliers. They too must check the output from their operation to ensure process soundness. More will be said in a later chapter as to how to handle material that has been identified as a defect.

The third inappropriate action practiced by each of these groups was to pass the defective part to the next operation without creating a remedial action. *Hold defective parts. Don't allow part or process continuation when a defect is observed.* (Do Hold.) Defective parts must be held and sequestered to prevent normal path continuation of the defect involved. Parts determined to contain a defect must be removed from the system, tagged, and repaired or scrapped, and the appropriate parts returned to the system as specified by the control plan in existence. These are steps that the supplier should mandate in their defined work process, control plan, and instructions. Any of the employees involved in the sequential operations performed could and should have prevented the sharp fin defect from being shipped to the store for sale.

2. A metal casting supplier provided damaged and flawed automotive and marine crankshaft castings to a machining operation after they were erroneously replaced back into the main line of production. These castings that had been removed and were determined to be defective did not follow the prescribed routing for defective components and should have been captured. These products passed through a machining operation without further difficulty and passed a final inspection before being assembled into automotive and water craft engines. A total of six engines failed and caused operator difficulty. The most serious effect was the stranding of a pleasure boat a mile offshore in Lake Erie.

3. A manufacturer of juvenile products provided one that required four D-sized batteries. They designed their own battery compartment in which it was possible to install the batteries incorrectly, with one or more upside down. This condition resulted in battery overcharge and overheating when one of the batteries was installed incorrectly. The overheating caused a battery explosion, with detrimental particles and vapors that damaged items within customer homes. (Walls, rugs, and curtains were damaged and required replacement.)

As can be observed, supplier problems can be created within one's own company by basic manufacturers, designers, and others who supply components, materials, or services for a product or service.

These are only a few of the myriad examples that can be recognized and attributed to inattentive suppliers. It is extremely important to know the strengths and weaknesses of those supplying parts or services. Each of these recognized problems resulted in monetary losses due to litigation, replacement, or customer relations activities. It is not unusual to ascribe supplier problems to government agencies, basic manufacturers, or others who supply components, materials, or services.

It is therefore important to know your suppliers. One of the methods of evaluating suppliers is to maintain a file or record on each unfavorable manufacturing condition that is experienced within your organization. Some businesses believe themselves to be too small to make this expenditure. I disagree with this hypothesis because it will result in future losses as process improvements will not be recognized quickly. A simple supplier discrepancy form or e-file that lists the supplier, the deficiency, and the corrective actions taken may provide adequate clues to prevent extended problem resolution times.

Before we go any further, allow me to ask a question in regard to a simple hypothetical circumstance. The conditions are as follows:

> Suppose you have a supplier of screws that are used as fasteners. These screws are installed into an automatic feeder that assembles them automatically onto an engine assembly. Unfortunately, when a mixed part is included with the specified fastener, it jams the machine and the production process is curtailed until the fault is corrected. Regrettably, this is a common occurrence with the fastener supplier. Your process has just encountered its second jam in two weeks.

If you are in charge, what do you do? How do you face it?

1. Clear the jam and continue production.
2. Ignore the problem because it only happens occasionally.
3. Find and save any out-of-specification parts for review.
4. Involve the quality supervisor and find out why it has happened.
5. Request someone, if not yourself, to determine why it has happened.
6. Have a quality or purchasing representative contact the supplier to voice concern and to display annoyance.
7. Specifically request that you be notified of the corrective actions taken to prevent recurrence.
8. Talk to a superior and request to have a supplier telephone call, hold a supplier meeting on your premises, visit the supplier, or conduct a mandatory supplier symposium at your site.

9. Remove the remainder of the supplied materials that were involved in the jam and reinspect them before continuing.
10. Return the balance of the material from this shipment to the supplier for a more thorough inspection. Or something else?

The examples in this book should allow you to judge your responses as you review the text. Hopefully, you will be counseled within the coming chapters as to the correct responses applicable to your needs.

Three types of supplier problems

As we continue, let's examine the severity of the problems and the effects of supplier actions on various operating systems. There are two types of problems that suppliers can cause. One is a minor difficulty that may cause a temporary condition—as, for example, the first problem below, which creates an inconvenience.

Inconvenience

A supplier was supplying machined components for an engine assembly line that required a lubricant coating to be applied to the parts before they were shipped. Over a period of time the oil dip degenerated, or other oil components were added to the approved dip mix. This did not present a problem during the hot months of summer. However, when the cold weather started, the dip congealed on the components and solidified somewhat, preventing the parts from being assembled because of the foreign material that coated them. The supplier did not immediately disclose that these oil dip changes had been made.

It is not uncommon to find that some suppliers will replace approved materials with less expensive material to improve their profit margin or in some cases change the process inadvertently or without permission. It is imperative, therefore, that suppliers be held to stringent pre-agreed and approved requirements and methods for their process control.

Assessed severity

Another more serious example follows that would be a major problem. The severity is assessed as a critical interruption of the process. This problem occurred when a fastener supplier was providing engine head cover fasteners that failed due to fatigue after they had been properly installed. The problem was discovered as the engines in storage were being shipped from the warehouse. A loader noticed that there were pieces of components on the warehouse floor after the engines were shipped. Upon further evaluation it was found that the engines had fractured and missing fasteners on the head covers. These engines had been inspected and approved for shipment during previous inspections and audits. Necessarily, this caused an immediate serious problem. The assembly line was shut down, completed engines in the warehouse were captured and engines in shipment were returned. The engine plant notified their customer, a truck assembly plant, to do hold their last shipments until verifications could be provided. An immediate recall of all shipped products to the assembly plant was initiated. And an investigation was conducted to determine the cause of the problem.

Luckily, an evaluation and inspection of the failed components held a clue that could not be ignored. It appears that the Tier 1 fastener supplier was experiencing machine problems. Their machine was double striking the components during forming, which induced a fatigue failure point. They had not followed the approved process in that they allowed the shipment of double-struck components. The supplier's admission that they had experienced and corrected a machining problem shortened the evaluation and correction process.

It was not inexpensive to do hold all of the engines on hand, those in shipment, and those at the engine assembly plant to ascertain the presence or lack of double-struck fasteners. Each engine was refurbished and inspected before they were released to the engine plant and the final customer.

Criminal activity

Another problem that can arise is a lack of controls that ensure the process productivity and the resultant profitability of the top and bottom line. A foundry facility operated many molding machines that produced cores that required gallons of chlorinated hydrocarbon fluid of the proper density, fluidity, and consistency to allow proper core and mold compaction. The fluid was provided to the customer in 55-gallon containers during weekly shipments. It was noticed that the inventory was inaccurate and weekly requirements increased unexpectedly.

When the process was reviewed it was found that a few customer and supplier employees colluded to create an illegal increase in their incomes. They were observed taking one-half of the supplied barrels, marked with yellow chalk, to the far end of the property and pouring the contents into a storm sewer (Figure 2.3). Luckily, these drums contained only water, so they didn't contaminate the sewer system. This increase in hydraulic fluid usage created a powerful increase in the operational costs of the customer. It was found that these dishonest employees returned the empty barrels to the supplier to be refilled with water, which could again be spilled into the storm sewer. It appeared that the profits obtained due to the theft by supplying water in place of the chlorinated hydrocarbon were being split by those involved. This was first discovered during an auditing operation

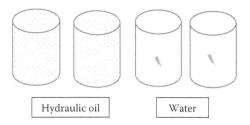

| Hydraulic oil | | Water |

Figure 2.3 Specious composition of the delivered oil supply.

when the empty barrels were discovered to be not confined to a designated hazardous waste storage area. (The company was paying the supplier an exorbitant amount of money per gallon for the hydraulic fluid, which impacted the bottom line.)

Now, each of the above types of supplier-induced problems should be of value to the reader. These fallacious conditions required that corrective actions be taken by the original supplier and the customer to prevent recurrence. Unfortunately, it is necessary to verify, as well as trust, that the systems in place are working in a controlled manner. If systems can be compromised, they will be. Audit tools can aid the supplier in identifying supplier problem areas. These tools will be provided in a later chapter.

Immediate revision to the supplier control plan, work instructions, and failure mode and effects analysis* (FMEA) were mandated for the fastener supplier. Not to be one-sided about this, the engine assembly plant and the foundry operation were also required to modify their control plan, work instructions, and FMEA, as they too passed along defects to their customer or allowed improper practices. But more about this later.

Summary

One of the problems, then, is the lack of recognition that most suppliers can be a viable partner in any endeavor, and that their actions can have a direct and an indirect effect on profits. This partnership can be accomplished through supplier development and enhancement. Most suppliers are honest business establishments that seek the respect of their customers. However, there are situations where a few dishonest suppliers or individuals can taint the entire supply system. Some quality and purchasing managers have found that treating qualified suppliers as business partners rather than as adversaries produces the best results. This relationship between the customer and their suppliers is achieved by

* The information within does not describe those methods or procedures of design FMEA, process FMEA, or control plans that may be mentioned. The operational description of these methods can be found in the *Quality Control Handbook* by J. M. Juran, published by McGraw Hill, Inc., New York.

mentoring those suppliers in the practice of self-imposed and mandated enhancement corrections. Suppliers not held accountable will continue to provide services and products that are not acceptable. A series of proven best practices, processes, systems, and tools were accumulated over a period of years by resolving supplier problems in the automotive, chemical, electronic, gear, engine, casting, assembly, forming, clothing, juvenile products, and toy industries. These practices will be explained in the following chapters. Unfortunately, it is imperative that new or inexperienced suppliers be scrutinized at frequent intervals at the beginning of their service to ensure that quality and quantity requirements are being met.

This chapter also describes three specific operations that can be used to develop both internal and external suppliers. This is a philosophy of holding employees responsible and accountable for performing three functions: the act of conducting a cursory inspection of items before and after their operation, the requirement to perform the work correctly, and the sequestration and holding of material that may be questionable without sending it to the next customer. These are some of the methods that can be used to improve profitability through the application of supplier modification. These requirements were formulated over many years of mentoring many different suppliers in various industries.

The next chapter describes the types of suppliers that may be encountered by a customer. It gives examples of supplier traits that focus attention on differences that are to be considered when dealing with suppliers. The chapter relates instances of supplier inaction or impertinence that resulted in inefficiency and dollar loss. It gives further examples of destructive activities from unfit suppliers. Finally, the chapter introduces some supplier control and improvement concepts to be studied. It describes the differences and similarities in their credentials when interacting and taking corrective steps with the customer.

chapter three

Need for enhancement

The previous chapter presented the results of a study on supplier improvement. It discussed those unsatisfactory conditions that were found in comparison with those conditions that are to be desired. In addition, it proposed actions where necessary to correct the supplier responses to problem conditions.

It is necessary to discuss the study history first for relevance. Most of the time a manufacturing or service organization may not pay particular attention to their suppliers' needs until a problem is encountered. However, once a problem is encountered, a supplier may be blamed for an activity that has caused the problem. There seems to be a human tendency within some firms or organizations to believe that an external supplier is to be automatically ascribed blame. This is not an efficient way of resolving problems. Qualified suppliers should be recognized as business partners to generate the best results. Now, that being said, there are many suppliers that may require strict scrutiny in all of their services. Both of these types of suppliers will be discussed within the framework of this book. The best practices that can be gleaned from the best suppliers should be applied to improve the less satisfactory suppliers.

This chapter provides the basis for the need to address supplier enhancement. A preliminary study was conducted with an automotive supplier base to evaluate the need to improve the customer–supplier relationship. The investigation involved looking at service conditions during discussions and evaluations in order to act on supplier problems that were experienced. It addressed the conditions that were found to exist and proposed corrective actions.

Best and worst practices

The study included an investigation to determine the differences between the best and worst practices involved in supplier problem solving. In general, the best practices revealed simple and clear communication of the problem at hand. The supplier expertise in handling problems was related to their experience and the general size of their business. Suppliers to large organizations were more apt to handle not only the evaluation but also the corrective actions necessary to eliminate any recurrence. In general, the following conditions were found to be the most useful to expedite supplier quality problems.

Best practices

1. A concise definition of the problem
2. A photograph or sketch that illustrated the problem
3. An agreed visualization and recognition of the problem
4. An understanding of the date code or identification used
5. A certified measurement system
6. The use of variable data to address the problem
7. A compilation of the available data and information
8. A record and availability of the information gathered
9. Updated information available before meetings
10. Experience with basic problem-solving tools
11. Cooperation in the problem evaluation and solution

Worst practices

The poorest responses from suppliers was found to be from inexperienced or newly acquired quality representatives at small firms. There was commonly a lack of concise job experience and instructions for supplier employees. This included the lack of instructions to do check, don't create, or do hold materials in use in their facilities. (This philosophy requires employees to visually peruse incoming material [Do Check], to make only acceptable products [Don't Create], and to seize all defective products before they are forwarded for additional processing [Do Hold]). And finally, some of the quality representatives were lacking in basic problem-solving experience. This lack of experience was determined to be the biggest disadvantage assignable to the supplier. This deficiency also showed itself in the apparent lack of response in determining and executing corrective actions. It was construed that the reason there was little or no response was that these suppliers lacked an understanding of the importance of determining the root cause of the problem or providing a corrective action. So, if no root cause was established, no solution to its cause could be determined, and then no corrective action could be considered and applied. Improvements that are not applied can't be tested or verified.

These adverse conditions compounded the visualization of the problem because then the root cause was not recognized, the flaw was not prevented, and the defect was not captured at the supplier site. Further, since these actions were not taken, there could be no quality planning process in place to address the prevention of similar problems in the future.

Circumstances found

An evaluation of a series of supplier problems revealed that there were numerous circumstances where improvements were required. Some suppliers thought that it was not apropos to deal with information requests in a timely manner, if at all. Some suppliers resisted the sending of information on a timely basis, which affects the understanding and discussion of the materials. These supplier attitudes are not desirable.

What should have been present was the timely provision of the information that was requested, so that it could be reviewed prior to the next scheduled meeting. This would have allowed everyone to prepare for the meeting and to plan ahead to acquire other relevant information by request. Useful conference time was also wasted reporting conditions that should have been readily available.

More mistakes were included due to misunderstanding the nature of the problem, a lack of understanding of the materials being discussed, and the acceptance of partial information as a condition of certainty. A sample size of one is totally inappropriate and useless for developing or concluding a study.

A common vision and understanding of the problem—a realization that sketches and pictures of the flawed conditions were critical and that complete information about the problem was required—were woefully missing.

The finding that incomplete and poor problem definitions were accompanied by weak measurement systems, using unconnected rather than variable data, contributed further weaknesses because of the suppliers' lack of problem-solving understanding.

Most analyses lacked a strong definition of the problem and did not use identified data code and pattern serial information with the application of basic problem-solving experience. These attributes are necessary and desirable features to be used in conjunction with the skills necessary to solve rudimentary problems. Unfortunately, some suppliers lacked the skills to provide basic informational devices such as photographs, sketches, charts, or an analysis that would help to expedite problem solutions. They also may have felt uncomfortable in providing a description of the work that was to be provided in specific operations.

What should have been found was the supplier's cooperation in providing all photographs, sketches, charts, analyses, and job work instructions as may have been required. Whatever the reason, the supplier must be made to understand the importance of sharing this information to accelerate the solution of the problem.

Most important action

After a supplier problem is recognized, the most important action is to ensure the containment of any defective, questionable, or unapproved materials to prevent *quality spills*. (A quality spill is the release of product to a customer who will be adversely affected by the nonconforming material. This might also be referred to as impounding defective products, which might even include a product recall from the final customer.) This sequestration provides the containment of the entire supply and distribution pipeline from the supplier through all phases of manufacture, storage, or transportation. The containment helps to prevent the flawed product from affecting the final customer. This first action must be taken before any supplier quality-solving activity is started. Then, certainly, it is necessary for all suppliers to apply the same philosophy, so that their customers are not affected by adversely problematic products.

Established practices

Because of the weakness observed in some suppliers, it was desirable to establish a basic system in an attempt to improve the supply system. Practices were established to address the suggested improvements in five areas for all suppliers:

1. Suppliers were to be held accountable when their first inconsistency was discovered and were to react immediately with information and proposed corrective action.
2. Suppliers were requested to use acceptable problem-solving tools to aid their problem evaluations and to make results available in a timely manner.
3. Suppliers were required to repair, refresh, refurbish, or replace items rejected at the customer plant. (These are materials that were found to be unsatisfactory upon arrival at the customer.)
4. The use of a problem data sheet was assigned to capture relevant definitions and data at the onset of problem discussion. This sheet was to contain inputs from both the supplier and the customer.
5. The problem data sheet was to be filled out and provided to all meeting participants an hour before the next call or meeting.

Initially, it was necessary to contact all suppliers to enlist their cooperation as the corrective system was established. Once in place, the requirement of the practices was included as part of the pre-award meeting that established the supplier contract.

Conditions eliminated or controlled

The following conditions were the result.

- Relevant information was available for meetings.

- Communication gaps became known and were overcome.
- Supplier weaknesses became apparent for correction.
- Problem-solving tools were identified and required.
- Positive meeting results were improved and progressive solutions were expedited.

Problems encountered

Whenever a new system is introduced, it may experience new types of problems that were not experienced before the new system was introduced. Some suppliers did not expect such concentrated scrutiny when they were notified of each disturbance. In retrospect, they appeared in awe that small simple problems were attracting significant interest from the customer.

Many suppliers, upon being contacted, listened to the problem as described by the customer and stated that they were not aware that the condition was a major problem and that they would have to call back after they had conducted an investigation. They were surprised that basic analysis tools such as histograms, concentration diagrams, analyses, flow plans, or other evaluation techniques were required. If they called back without any of these instruments to discuss the problem as requested, an agreement was reached as to their necessity. Apparently, in some cases the supplier purchasing managers inadvertently forgot to pass on the requirement that they had agreed to. This change was then rapidly initiated.

It was found that the use of problem analysis was not widely understood or correctly applied to supplier enhancement. For example, a problem was created when a wrong part caused a production line feeder to jam. One of the first examples of an incomplete analysis was the input of a supplier that listed the following steps:

Problem: Part caused a jam of the production line feeder.

1. Wrong part was included in the shipment.
2. Wrong part was not detected before shipment.
3. Foreign material was not controlled for this shipment.
4. Some assignable cause is present.
5. Quality planning to prevent mixed parts is inadequate.

Finding: There was a lack of planning to prevent mixed parts.

Recommendation: Evaluate process to find wrong parts. The supervisor will reinstruct his employees.

Now, as you can observe, this exploration did not determine a definitive root cause of the problem. In addition, it did not focus attention on the prevention methods that were needed to prevent recurrence. There is a lot more required in an evaluation to enable the system to prevent repetition and to become effective.

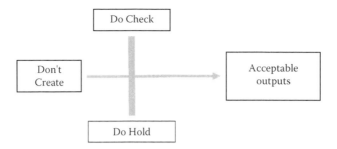

Figure 3.1 Trilogy of supplier enhancement.

An origin assessment for supplier problems was created, as shown in Figure 3.1. It addresses the three proposals that must be considered in improving supplier development.

The cycle of maintaining supplier consistency and improvement will be addressed more fully in a later chapter, where we address the problem as experienced between the customer and the supplier.

Summary

This chapter reviewed the results of a supplier enhancement study. It discussed what was found in comparison to what should have been found and proposed corrective actions that would enhance the supplier performance. In addition, the chapter stated the most important item when finding a supplier defect. (Corrective actions must be taken to prevent a problem recurrence.) Furthermore, it is important to determine if it will affect your customer. If it does, you must initiate immediate corralling action to prevent the flawed material or service from reaching your intermediate and or final customer.

The supplier–customer relationship is a valuable asset that should be engendered. This relationship provides cooperation that can be used to improve quality output as well as other necessary improvements to increase profitability.

The next chapter discusses the need for supplier improvements. It presents a sample of historical problems created by suppliers. It relates the applicability of collecting data to pinpointing problems and the methods that will resolve those problems. The chapter gives examples of supplier traits that focus attention on differences that are to be considered. The discourse relates instances of supplier inaction or suitability that result in inefficiencies and dollar loss. In addition, examples of the destructive activities of unfit suppliers are illustrated. The need for enhanced supplier control and improvement concepts is focused on with examples.

chapter four

Supplier credentials

The previous chapter dealt with the problems faced and lessons learned over many years with individual suppliers involved in many manufacturing or service endeavors. Most suppliers are honest and are willing to cooperate to address any problem that occurs. (This is said with tongue in cheek. It is true that if it does not increase their costs, create perceived hardships, or cause expensive revisions to their process, they will cooperate willingly.) But unless they are actively involved in all types of supplier improvement actions they will not achieve the status of having enhanced suppliers.

This chapter deals with the two taxonomies of suppliers that are experienced in everyday industry. It will describe the differences and similarities that are involved in their actions, interactions, and corrective steps taken with their customers. Descriptions of the types of supplier that may be encountered and their ability to conform to international standards are related. Examples of supplier difficulties are discussed. The difficulties must be addressed with the application of corrective actions to the design, the process, and the work instructions that are practiced.

Types of suppliers

The first type of supplier is internal. That is, the supplier can be anyone within the customer's organization that has a hand in providing a service or product to help them fulfill their activity. It can involve individuals from the mail room to the boardroom. Anyone that can influence an activity within the company that has a direct or indirect influence on company performance can be considered an internal supplier. Generally, these departmental suppliers lack the individual credentials required as an outside establishment. That is, they are subject to the norms of the institution in which they work.

The second type of supplier are those that are not within the organization or company. They are independent entities that provide the customer with a service or product, generally on a work order or bid-approved basis. These suppliers are generally referred to as *Tier 1* or *Tier 2* suppliers.

The difference between a Tier 1 and a Tier 2 supplier is their location in the supply chain. Tier 1 suppliers provide parts directly to the customer, whereas Tier 2 suppliers provide a product or service to a company's Tier 1 supplier. The Tier 1 supplier then adjusts, assembles, molds, mixes, coats, forms, trims, or otherwise processes the product that they received before shipping it to the customer. The supply chain is depicted in Figure 4.1.

Tier 2   Tier 1   Customer

Figure 4.1 Tier location in a supply chain.

Mine coal   Inspect and sort   Ship by rail   Convert to electricity   Provides heat and light

Figure 4.2 Tier flow path from coal mine to home service.

Now, there is an undeniable similarity in the definition and recognition of internal and external suppliers. Necessarily, what defines both is their contribution of a service, product, or input to the next customer. You can also appreciate that the Tier 1 supplier is the customer of the Tier 2 supplier.

It is also recognized that there may be more than two tiers of suppliers to the final customer. It is important, then, to realize that using a Tier 1 supplier to mentor and hold responsible their suppliers is necessary for total supplier accountability. This is why Tier 1 suppliers should be treated as business partners, because they can impact the quality and performance of the product or service that is provided to the final customer.

The tier flow plan in Figure 4.2 shows various processes or services provided by the tier suppliers. The Tier 4 supplier mines the coal. The Tier 3 supplier removes large rocks from the coal. The Tier 2 supplier moves the raw material to the power plant. The Tier 1 supplier converts it to energy, and the final customer uses the energy to create heat or light. Most, many, or few of the suppliers in this chain may have the necessary control tools in place to be certified by an international organization. The more organizations in a chain that are qualified by certification, the better will be the final results in terms of control, responsiveness, and cost.

Consider for a minute the requirements that are involved in the process. The mine must provide the coal required, whether it be a soft coal or an anthracite coal. In China, the Tier 3 supplier assigns many employees to stand adjacent to a conveyor belt to pick out the large rocks from the coal as it was being delivered to the coal storage pile. The Tier 2 supplier has to provide dedicated clean container cars free from contaminants that could damage a power plant's equipment. The Tier 1 supplier must provide electricity in a specified generation range to ensure that a consistent product is produced. Each of these operations or services is important to the success of the final customer.

If the wrong coal is supplied, it may result in excessive smoke or combustion. If large rocks are not removed from the coal on the conveyor, the power plant may suffer damaged equipment. If dedicated rail cars are not available, shipments might be curtailed, delayed, or contaminated. If incorrect electrical properties are generated, the voltage, the amperage, and the power factor relationship between them could result in final customer dissatisfaction. So, then, if each customer requires that their suppliers take irreversible corrective action whenever a problem has been experienced, the system will morph into a more effective one. So, the importance of a competent tiered supplier system becomes readily apparent.

Supplier traits

Giving consideration to selecting and keeping a supplier is paramount to the smooth operation of a company. Internal suppliers can be nudged into compliance by the effective use of a management system that utilizes group and individual evaluations or managerial oversight.

External or outside suppliers may require more stringent control if they are to be a viable source. This is why it is advisable to use outside sources that have been certified by outside criteria that involve international standards. This may not always be possible as the product may be supplied by a "mom and pop" shop that lacks the expertise and functional management. This book is not primarily concerned with those suppliers that have accreditation through outside audits by a regulating body such as CFR, QS, or ISO. (These regulating bodies have specifications or requirements that must be maintained.) Their systems require a response and a corrective action to be taken whenever a discrepancy is noted. Rather, this book is dedicated to correcting those suppliers that lack proper procedural application or those that require additional help to fulfill their roles. These providers may include certified suppliers that are not cognizant of all the current corrective action tools to be used. Or it can be applied to weaker suppliers that have not yet enhanced their problem-solving skills.

Narrative on suppliers

Suppliers are not a necessary evil. They are valuable partners in that they provide a product or service that is required by others. They cannot be judged simply on their appearance. Take, for example, the following illustration (Figure 4.3).

Looking at the photograph, it might be concluded that they are all the same. In actuality they are not. Each of the coins shown in Figure 4.3 is of equal value to a purchaser. However, they are not identical. The upper row is distorted slightly to make them larger, to allow the picture to be taken

Figure 4.3 Six American coins.

with minimum light reflection. The coins have various dates and there are indications of different locations to indicate where they were minted. They have different shading, hues, and clarity. This is not dissimilar to the comparison of suppliers. Each one has their own specific strength or weakness that determines their value. It is up to the customer to weed out the weak or defective suppliers from their system to ensure competiveness.

Suppliers needing improvement

Some suppliers are distinctly defective, although they still function as valuable suppliers to others. The picture in Figure 4.4 is akin to a defective supplier. Its imperfection does not show readily. Its distress may not show until examined by a potential user. Close observation of suppliers is required also.

Figure 4.4 Misplaced die strike.

Figure 4.5 A correctly minted coin.

The coin in question is that shown at the bottom right of Figure 4.3 and then alone in Figure 4.4. It has what appears to be a misplaced strike by the coin die. The words "IN GOD WE TRUST" have been imperfectly minted and appear to be out of position. The letter I in "IN" and the word "WE" are barely perceptible. The letters T and Y in "LIBERTY" are truncated and not aligned along the edge of the coin with the rest of the letters. The imperfections are not due to wear as the periphery of the coin is raised higher than the print. It is not comparable to other similar coins where these deviances are not present, as in Figure 4.5.

What could potentially have caused this problem? It could have been an inspection oversight. It could have occurred due to a machine malfunction. It could have been done on purpose. There are sundry reasons why this could have happened. The important point is that if it occurred at all there is the potential that it will occur again, unless permanent corrective actions are taken. This is not dissimilar to the rationale of correcting supplier problems. Unless corrective changes are made the situation can cause problems again. These corrective actions will be discussed more fully in the chapters to come.

Before we continue, be aware again that there may be a certain limited number of suppliers that do not care about your success. They are strictly concerned with their own and will create problems that will interfere with your goals. Generally, they will not survive in the supply system for any extended period once their deficiencies are recognized. Here are some samples of the lack of honesty of suppliers resulting in dollar losses and, in some cases, unsafe conditions for product users.

Red dye problem

A toy company employed many offshore companies to provide them with the manpower to produce toys to be sold the entire world over. To be viable the toy company had to comply with U.S. and international standards to ensure that they were complying with all applicable regulations. The company was required to meet stringent requirements to supply items to toddlers less than three years old. These included the omission of small parts that could be ingested, the use of components with suitable breakaway force, and the use of paints that were not toxic. These three requirements were mandatory to comply with CFR regulations and other worldwide safety standards.

When generating an offshore supplier base, one of the toy company's requirements was to approve a supplier with an acceptable preproduction approval process. Before the final contract approvals were given, both customer and supplier companies agreed that there would be no deviation from an approved process that specified the location, method of production, suppliers, materials, and process controls. This is critical when dealing with international standards.

Unfortunately, the supplying company decided that they could save a significant amount of money if they changed one of the elements of the toy. They decided that they could achieve the bright red color that had been approved for the toy with a different base chemical. So, inadvertently or purposefully, they used cadmium as their base for the red color. The color produced was very close to the red color that had been approved. However, once it was noticed, a sample was analyzed in the laboratory facilities and was found to have a restricted paint base (cadmium). Production was immediately in jeopardy. A product recall was initiated. All customer orders were cancelled and the new product was not released. The vast amount of inventory on hand, in transit, and at the offshore provider were sequestered and scrapped. This caused a major problem in more ways than one.

The supplier was not paid because they didn't provide the materials as ordered. The new product release was scuttled along with all of the ancillary advertising that had been prepared. Sales were lost because the store shelves were not supplied with a product that was to fill a market niche. Invaluable engineering and marketing time invested in the new product was lost.

A lesson to be learned from this is to demand not only that locations, materials, methods, suppliers, and process controls be agreed on but that they also be in compliance. For this reason it is necessary to acquire documentation and certification that locks in a process description before granting a production release (i.e., a preproduction approval process to qualify all of the important aspects of production). Unauthorized changes are not to be permitted. Nothing, absolutely nothing, is to be changed without the prior agreement of both the customer and the supplier once production approval is granted. All agreements must be mandated to ensure compliance to regulations.

Disappearing materials

The next example of a corrupt supplier involved a private contractor that supplied raw scrap metal materials to an automotive foundry. This service was conducted under contract to deliver scrap metal by truck from their scrapyard to the foundry operation. When the truck entered the property it was to report to a scale area, where the truck and the inherent load were weighed on a railroad car scale. After weighing it was to proceed to a designated storage dump area, where it would dump the material load. After unloading it was to return to the railroad scale area and be reweighed. This was believed to be a fair and equitable method for determining payment between both parties for supplying the scrap materials to be delivered for the melting operation.

A problem was noticed in the inventory of the scrap material accumulating in the storage area. It was difficult to explain the weight variance between what was available and what should have been available because the weight of the piled material could only be estimated. The driver almost always stopped in the plant cafeteria for a cup of coffee each time that he delivered a load of scrap metal. Over the next few weeks it was noticed that the truck left a puddle or a trail every time it unloaded its contents. The truck always weighed approximately the same when it left, so it was not possible that the driver was only unloading a half load.

Further investigation was conducted after the realization that the scrap metal pile was deficient and that the dump area was always very wet after a delivery. Closer observation of the delivery operation was conducted next to the truck as soon as the driver entered the cafeteria for his coffee. It was found that there was a stream of water spilling from the front of the truck bed, discharging many gallons of water. Apparently, there was a sealed storage hopper in the vehicle that could contain water without spillage until a valve was opened at the base of the tank.

Obviously, this appeared to account for the lack of steel and iron scrap in the raw material pile. This also answered why the truck weighed the same when it was reweighed after it delivered the load. Once emptied of water the truck weight remained within acceptable limits. The customer plant was paying scrap prices for water. Needless to say, it was embarrassing to report the situation to the plant manager. I don't know how long this scam had been perpetrated.

Most suppliers are honorable and should be considered to be valuable partners. However, every once in a while there are dishonest suppliers who take advantage of circumstances to profit inappropriately. Conducting audits and checks at regular intervals can help to pinpoint those practices that are detrimental to your operation and profitability.

Supplier negligence

Flint is a city north of Detroit, Michigan, in the United States. It has a tax base that was decimated by a loss of funding due to the decline of the automotive and other industries. One of the corrective actions taken by Flint's government was to reassign their acquisition of assets for the city. One of the assets that was scrutinized was the source of the city's water supply.

It appears that Flint was currently receiving their water from a nearby city but considered the acquisition costs to be excessive. To save costs, it decided to reinitiate an old system that had supplied water from the Flint River in previous years.

Water sources used for human consumption are regulated by the EPA and other local environmental agencies. In addition, they are generally checked for contaminants on a regular basis to ensure public safety. It was reported that the government agencies found that the water system was contaminated and that it contained an excess of lead particulate, which made it unsafe for human consumption. It also appears that the testing agency notified the local authorities of the dangerous conditions and then ignored enforcing rules to correct the unsafe water supply. In addition, the local agencies and the water supplier did not respond to the dire warning about the biological hazard that was present.

This condition of neglect and inaction created unsafe drinking water to be provided to a community for more than a few years. No one reacted to the reportedly high lead levels in the municipal water supply. Many lives were affected and many suffered adverse effects because of the duration of the contamination. Consequently, thousands of people suffered from various degrees of lead poisoning or contact.

So, it is imperative that actions be taken whenever a process is determined to be defunct. It is also necessary to ensure that corrective actions are taken in a timely manner to correct deviant conditions. These corrective actions and process changes should be mandated, irrespective of whether the violation has been found in a "mom and pop" shop, a manufacturer, or a government organization. Corrective actions must be applied whenever a defect is recognized to ensure continued success and to prevent recurrences.

Now, all three of the above suppliers committed actions that could have been prevented or reduced by more stringent supplier audits and checks to ensure enhancement methods were practiced. These methods will be discussed in a later chapter of this book.

Summary

This chapter viewed supplier differences. It described the types of suppliers that may be encountered and the importance of, whenever possible, using suppliers that are certified as conforming to international standards. In the discussion of supplier examples it provided an introduction to why it is necessary to establish preapproved processes and to conduct continual random audits to ensure continuity. It also introduces the need for supplier enhancement tools, which will be provided in a later chapter. Incorporating a suitable supplier selection and improvement system will ensure the continuity of smooth operations.

The next chapter reviews those supplier attributes that must necessarily be considered before a supplier is chosen for their involvement.

chapter five

Supplier elements

In the previous chapter we discussed the need to review, select, and improve the suppliers that we utilize. These suppliers could be internal or external. They could range from "mom and pop" shops to giant industrial manufacturers or service suppliers. All types should be held accountable for improvement once a problem is recognized. To hold them responsible it is first necessary to establish a selection or evaluation process to ensure full agreement that important provider criteria are fulfilled. Then it is necessary to ensure that a corrective follow-up action is established.

This chapter focuses on the supplier traits and practices necessary for success as experienced by the customer. Certainly, the more important the contribution of the supplier, the more critical their cooperation and evaluation. By all means, if their failure can create a major production or service disruption, their competency must be fully scrutinized.

Maintaining and enhancing quality suppliers is akin to keeping correct dental hygiene. If proper practices are employed and maintained you should not experience any decay or extraneous costs. Helping to maintain and enhance suppliers is a symbiotic relationship in which both can benefit. This is applicable to both internal and external supplier bases.

Generally, problems caused by internal suppliers can be more easily handled by the local management team. Internal problems are more simply corrected by improvements to job instructions, procedures, and work rules. Normally, many deviant or unproductive practices can be recognized and dealt with by applying a corrective action system. That is, every time a problem is recognized, there should be a verifiable action taken to prevent it from occurring again. Included in the corrective action philosophy is the requirement to perform work that is conducted with specified tools, methods, actions, and motions to ensure success. These are important considerations when considering small manufacturers, services, or specialized "mom and pop" shops. Proper training is of the essence to ensure quality.

Focus of supplier traits

Assuredly, if suppliers can cause unwanted service interruptions due to process, capacity, or delivery problems, their attributes need to be scrutinized most thoroughly. Some of the important attributes are as follows:

- Type of work: If a supplier provides a service such as janitorial work it would not be required to subject them to intense scrutiny other than for safety and security concerns. Economic considerations could be applied.
- Supplier history: If suppliers provide a manufactured product, it would be advantageous to ascertain their competency by reviewing at a minimum their quality performance and capability records. The following chart (Figure 5.1) illustrates what a customer found after summarizing the defects that were created within a designated period of time at their automotive manufacturing facility.

  It can be observed that there is a treasure trove of information in the defect chart. These automotive suppliers have a propensity to create assembly and machining problems. In addition, there should be immediate concern that poor communication is rampant. What is the proposed supplier's history in these areas? It is important to find out so that individual problem areas can be eliminated.
- Supplier capability: Has the supplier the capability to supply the specialized product that they offer? If they are a screw, nail, or fastener supplier, can they meet the shipping requirements on time every time? What are their backup plans in the event of a fire or other emergency in their plant? It is important to determine whom they rely on for their inputs and what assurances their Tier 2 supplier has given them of their service. It behooves you to ascertain if they have ever notified any of their customers about a supply problem. If they indicate yes or no find out the circumstances.
- Supplier accreditation: Is the supplier certified to a national or international standard? If they are, cherish their work, as they will have

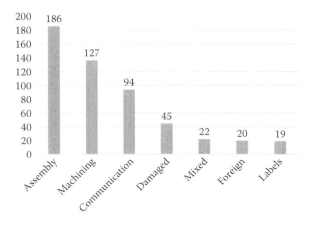

Figure 5.1 Supplier defects over time.

in place a lot of activities that will alleviate some of your concerns. This is not to say that they will not ever ship you a defective product, because there is no such thing as a perfect system. However, it most likely indicates that they have systems in place to handle deviations and the means to apply irreversible corrective actions. If unaccredited, mentor them and insist that it is a requirement of the supply system.

- Supplier skills: If the supplier is not certified or accredited to a national or international standard then there is much work to be done to help them become successful as a competent provider. The use of supplier enhancement tools should be applied, as will be discussed in a later chapter.

- Design failure mode and effect analysis (DFMEA)*: This type of tool will not be in use by the supplier unless the customer and the supplier interact throughout the design and planning stage of a product. It is a tool that is used at the beginning of a project that attempts to garner all of the pertinent information to prevent problems due to the product's design. For example, some automotive companies construct a DFMEA for all engine case and head components that will be manufactured in a new engine. A problem that required a DFMEA revision is presented in Figure 5.2, which shows a chip that was machined from the surface and which remained in the head casting after machining. Chips of this type can disengage and damage the engine. This consideration should be added to a revised DFMEA to prevent this condition from occurring again. This type of tool, then, should be applied wherever possible when the design of a product is being planned. You should expect your suppliers to apply this in preplanning or revise their existing DFMEA, PFMEA, and control plan as required.

- Process failure mode and effect analysis (PFMEA)[†]: This tool is similar to DFMEA in that it makes provisions for the process to be developed. It is dissimilar from DFMEA in that it does not primarily consider the design of the product but only the process that is to be established. A PFMEA contains a list of steps or operations with a list of potential causes or an evaluation of their effect on the quality of the product. A PFMEA is used as a tool to determine an operational score to indicate relevance, and it is used to rescore those problem operations after a corrective action has been completed.

---

* The information within does not describe those methods or procedures of DFMEA, PFMEA, or control plans that may be mentioned. The operational description of these methods can be accessed in the *Quality Control Handbook* by J. M. Juran, published by McGraw Hill, Inc., New York.
[†] The information within does not describe those methods or procedures of the PFMEA that may be mentioned. The operational description of these methods can be accessed in the *Quality Control Handbook* by J. M. Juran, published by McGraw Hill, Inc., New York.

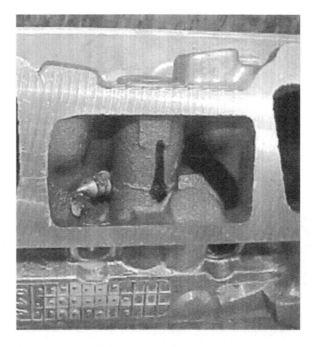

Figure 5.2 A foreign material chip retained in casting.

The analysis addresses process issues and is to be a limiting document in that what is specified on the document is the process that will be employed. It reflects a determination on the effects of variables on the basis of severity, incidence, and recognition. A PFMEA also lists the means of methods, controls, audits, and inspections that are to be conducted and controlled at specified times. It is a powerful roadmap as well as a prevention and correction tool. Every supplier must apply this tool and put it into effect.

Process control plans*:

These are the plans that have been specified to control the process. The control activities are listed on the PFMEA to specify the activity that will be used to monitor or eliminate the potential causes of problems that are identified by the PFMEA. These process control plans are important in that they actually define the circumstance that will be encountered in the production environment. Each process control plan may enlist other meaningful activities.

---

* The information within does not describe the methods or procedures of control plans mentioned. The operational description can be accessed in the *Quality Control Handbook* by J. M. Juran, published by McGraw Hill, Inc., New York.

- Procedures: All appropriate action descriptions should be present and available to the supplier plant population at the supplier work-station location. These are special procedures that should be applicable and audited from time to time to ensure compliance to good manufacturing practices. These procedures include, but are not limited to, first-piece inspections, continuous inspections, audits, gauge control, tagging, lockbox use, process flow path applications, storage, shipping requirements, reporting, and other follow-up processes. When auditing a supplier, either by self-inspection, visit to their plant, or verbally, it is necessary to discern if these appropriate procedures are in effect and being corrected as problems occur.
- Work instructions: These are used to specify the type of work that is to be performed. Sometimes you can aid a supplier in identifying one of a series of machining operations conducted on similar machines that is creating problems. It is often advisable to add a small paint dot in an obscure area that will not be apparent on the finished product to identify a machine, a location, or an operator difference. (The secure paint dot location should be chosen by the customer to avoid complaint.) Also, posting job instructions to supplement procedures should not be overlooked. These instructions should contain a listing of the tools, methods, sequence, special instructions, and the required safety equipment to be used. Any deviation in work instructions is to be observed and corrected as it may affect the quality or output. The supplier should provide the customer, via the PFMEA and a control plan, those improvements that have been established to prevent a recurrence of the problem.
- Capability: The supplier must be capable of supplying the intended product at the correct time in the correct amounts, irrespective of outside influences. To accomplish this task it is not necessary to apply statistical quality control to every process in the facility. It does help, however, to insist on statistical process control or certification for those variables that are critical to the operation of the final product. For example, the operation of a balancer device that is used on engines, both for racing and industrial applications, must reduce engine vibration while running. This device has a very simple construction (Figure 5.3). It is composed of an outside shell that consists of an outside container and a cover with a machined flywheel that is positioned within the shell, which is welded into a single assembly. Before assembly and welding, a specific amount of a very viscous liquid is inserted to cover the empty area between the flywheel and the housing. The viscosity of the liquid acts as a damper on the rotating flywheel contained within the housing shell. It was found that a supplier's viscosity quality checks were not accurate and were causing the dampers to function inefficiently. (The supplier was providing

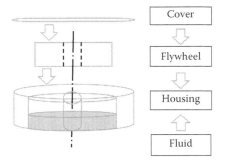

Figure 5.3 A damper balancer assembly.

an incorrect fluid viscosity.) To overcome this problem the supplier was required to submit the 55-gallon drum shipments to an outside laboratory for analysis. The results of the outside lab checks certified that the viscous material met the design and process requirements.

The use of the three major systems—DFMEA, PFMEA, and control plans—will enable you to enhance your supplier base if they are not already adequately incorporated into their operations. These tools will be instrumental in aiding you to create an information and improvement system that will result in better supplier quality. It is not unrealistic to insist that a potential supplier provide an example of their current systems for any joint venture on a proposed project. These systems should be a precursor to awarding any supplier with a contract if they will affect your manufacturing operation and productivity.

Summary

This chapter listed three important supplier attributes that should be available at the supplier location. The DFMEA, PFMEA, and control plans that they may have will be instrumental in their improvement and development. Generally, all certified suppliers should have these tools already in place. If considering a supplier that does not possess these attributes, it will be necessary to help them develop their methods, with your insistence and guidance, to achieve certification.

The next chapter deals with the means to resolve supplier problems, depending on the circumstances encountered. It describes aspects of daily meetings, scheduled supplier conversations, plant visits, and the use of supplier symposiums to establish better control of problem-solving activities. It introduces audits and other corrective actions to eliminate problem recurrences.

chapter six

Course of action

In previous chapters, many characteristics that identified the *who, what, where, when, why,* and *how* requirements of suppliers was discussed. Included in the discussion was the relevance of important process tools that are to be instituted by each supplier in their design, process design, process control, and corrective action systems. In addition, Chapter 5 listed three important elements that must be available and practiced at the supplier location. The design, process, and control plan will be of significant help in improving supplier enhancement.

This chapter suggests an innovative technique for accepting supplied components or assemblies without any additional in-house inspection by the receiving location. It also specifies suggested employee actions to ensure product integrity. Included is an example of a problem resolution sheet that can be used to drive corrective answers. These actions are driven by conducting the meeting types described. The sheet designates how actions can be taken to resolve any supplier problems that may arise.

Incoming inspection elimination

It is proposed that to remain competitive innovative manufacturers must strive to create a culture that fosters supplier engagement and teamwork to achieve the greatest possible symbiotic relationship. This fostered relationship will encourage yet greater creativity and innovation, which will create a more prosperous business for both. Let's first address the feasibility of accepting a product without performing any incoming inspection. Figure 6.1 shows an example of a simplified process that reduces inspection and shipping costs for both the supplier and the customer. It is based on a requirement that whoever is involved in product service or manufacture be held responsible for employing a simple change in philosophy. Employees, whether internal or external, must be held responsible and accountable for their work.

The flow plan shown in Figure 6.1 reflects a current system where any deviant material is shipped back to the supplier, which incurs additional shipping, rework, and inspection charges. The proposed system eliminates the application of the formal customer receiving inspection and requires that the Tier 1 supplier use their representative to perform remedial repairs as may be required at the customer facility or offsite.

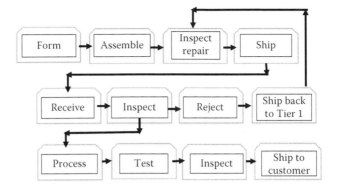

Figure 6.1 Hypothetical process flow plan.

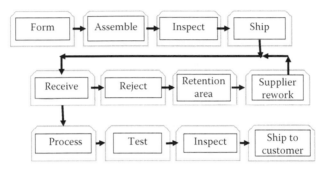

Figure 6.2 Proposed flow path.

(See Figure 6.2 for the revised flow path.) An example would be fin removal on plastic or metal cast parts such as toys, intake and exhaust manifolds, heads, cases, and pistons. This work could be performed at the customer site.

Whenever a defect is noted it should be sequestered and held so that it can't become engaged in the production process. It is easier to remove the defective part to a refurbishment area where it can be repaired than it is to require an assembly to be disassembled, repaired, and then reassembled.

Another example is a check for a missing component in a prepacked box of product from the supplier—for example, a missing screw in an assembled component. Under this flow plan and system, received components are immediately made available to the production line. At this point, the tenets of employee responsibility to peruse incoming materials, to capture and tag defective materials, and not to produce errant products takes action. Any deviant product is immediately removed from the process and is captured and held for a retention period until the supplier reworks or renovates the parts or assemblies into an acceptable condition.

This requires no additional shipping costs. These units missing one or more components could be refurbished at the customer location by the supplier's representative without requiring that the product be returned to the supplier facility, thus eliminating excessive shipping and handling costs.

At first you may be uncomfortable with the proposed system. However, both the supplier and the customer receive cost benefits due to reduced shipping. The suppliers improve their process controls because creating downtime at the customer line is very cost prohibitive (especially if the supplier is held accountable). Past experience indicates that they are rigorous in correcting their process controls and systems to prevent similar occurrences.

Employee responsibilities

Years ago it was ascribed as common knowledge that three 100% inspections were only rated to be about 80% effective. That is, if three persons stationed in sequence in a production line were assigned the tedious task of performing 100% inspections on the same rated parts they would only remove 80% of the defective items. Since then it has been recognized that the supplier should have a robust system in place that holds their employees responsible for the product that they form, assemble, or service. This has led to a philosophy that requires worker involvement and specifies the following:

1. *Do perform a cursory inspection* for obvious defects before and after your performed operation. (Do Check.)
2. *Don't create a defective product.* Take no action or inaction that would cause a defect to exist. Stop the process if a problem is observed. (Don't Create.)
3. *Sequester and tag defective items.* Use a lockbox or designated holding area to prevent part continuation. Don't allow a defective part or process to continue. (Do Hold.)

These three actions should be included in every manufacturer's internal and supplier instruction manuals. The person making the product should be responsible and accountable for the quality of the product that they service. If these procedures are established by the supplier then there is generally no need for the customer plant to perform any incoming inspection on the parts if they are not life-threatening.

I admit that there are some conditions that prevent this approach, but they are miniscule when compared to regularly manufactured product conditions. The savings in reduced shipping and inspection costs alone will impact the profitability of any enlightened organization.

Some suppliers and customers may be vehemently against adopting this system. As a supplier, I myself found that I was incorrect in resisting this activity as a provider to a large automotive customer. As a customer, I found that this system, once implemented, was useful and beneficial. Suppliers must be developed to agree to these provisions as they are beneficial to them and to their customer.

Of course there was a transition period and a learning curve for the supplier's manufacturing employees, but it was found to be a rapid transition. Any shipment found to be defective was either captured, held, and corrected by the supplier's representative in a retention area or returned to the supplier at their request. This protected the customer, provided quality components, and allowed the supplier to establish corrective actions at their base.

This method of reduced inspection and control should be considered for incorporation as it is an effective means of improving the supplier base. Once suppliers more fully realize that it is not permissible to supply defective items, they will become more aware of rejected products, which influences their bottom line. The profit motive is a strong influence that can be enforced by the application of contracts by the purchasing function at the customer location.

So, let's assume that there has been a purchasing and supplier agreement about the corrective actions to be taken in the event of problems before the supply contract is finalized. It should be required that the supplier assign a champion to handle problems for your facility. A champion is usually a quality engineer who is familiar with the tools that were mentioned in Chapter 5.

There are at least five activities that should be considered when approaching supplier-related problems. These are

1. Immediate contact
2. Conducting problem-solving meetings
3. Visits to the suppliers
4. Supplier symposiums
5. Problem resolution

In a previous chapter we posited a situation that required an evaluation of the actions to be taken when a problem arises. Hopefully, the following section will help to determine your course of action if the needs arise.

Immediate contact

I suggest to you that no problem is too small to be recognized and addressed. Obviously, it is imperative to address the problems that impact internal costs the most on a priority basis. Then, depending on the severity of the outcome, it behooves us to take actions to prevent recurrence. A good way to evaluate if a problem requires immediate attention is the application and use of a *Pareto listing*, where the most frequently occurring item is the most detrimental. But that may not provide you with the most supplier enhancements and cost savings unless it is plotted with dollar impact in mind. Rather, take each problem and assign a monetary value to the problem or interruption that it has caused. You will find that it is advantageous to act on the most deleterious conditions first.

Conversely, no problem is too small to be corrected. Once priorities are established it is necessary to address the high-cost items first. Then, without hesitation, it is necessary to find the time to notify the supplier of any problem that the facility may be experiencing. No matter how small the problem, the quality engineer or champion at the supplier facility must be notified. This allows them to consider the corrective actions to take to prevent another incident.

Now, it has been found that the best way to accomplish this is through a prearranged contact that has telephone availability. The use of texting for this purpose is not satisfactory. Human contact should be employed to notify them of the problem, to request their response as to what can be done and when it will be completed. This requires follow-up by someone within the customer organization. It certainly helps to utilize a service with audio and visual capabilities such as GoToMeeting, as the information and interactions can be enhanced.

Since most knowledgeable companies have some type of system to record and track supplier problems, this should not be an insurmountable problem. These types of systems will not be discussed here other than to indicate that any supplier problem is identified by identification number, supplier, date, problem description, number of occurrences, severity, and corrective actions taken to indicate the current status. It follows that these documents track the progress of the problem resolution from beginning to end. The system can also provide a history of repeat or similar problems and the irreversible corrective actions taken to prevent another occurring.

It is not unreasonable to expect daily or weekly telephone contact with the supplier's representative, depending on the severity of the problem. Follow-ups may also be required to ascertain additional information (this will be discussed later in the chapter), so require that a responsible contact be available to address your concerns immediately.

Conduct a series of meetings

Some problems are critical and require immediate attention from those managers and supervisors involved that are needed to rectify a serious condition. For example, a fastener that had previously attached a head cover to an engine suddenly began to fail after it was assembled. This condition jeopardized not only the manufacturing of the engines but also the operation of the truck assembly plant. Since this type of problem had no previous history and the failures affected operations significantly, it was mandatory that immediate and continuous contact be established to rectify the problem.

The engine plant quality manager notified the CEO, who was the supplier plant manager, and requested an immediate formation of a problem-solving team. The manager assigned himself, his quality manager, and a quality engineer to the task. The engine plant quality manager involved a quality engineer from the customer facility. The team agreed to have a daily call at 10 a.m. each day to resolve the difficulty. It was agreed that the quality engineers from both facilities would investigate from their respective positions and would act as record keepers for the meetings. A concrete meeting time commitment was established between supplier and customer.

During an investigation of this type it is necessary to have some experience in the problem-solving process. There should be a concise statement of the problem, an assigned scribe, and a roadmap to gather information. Capturing many samples of the defect is critical for assessing evaluations. The practice of providing a sample size of one is unacceptable.

Another important consideration is the severity of the problem to be addressed. In some cases a telephone call may suffice. In others a video conference might be opportune. Faxes, sketches, or photographs should be provided where possible. In other circumstances, it might be necessary to disperse a qualified representative to the supplier facility to encourage actions. This may be true whether the supplier is located a few miles away or somewhere else around the world. (This action depends on the severity of the problem experienced.)

Figures 6.3 through 6.5 indicate types of problem evaluation sheets that can be employed to capture relevant problem data. They can be used as the primary means of record keeping and information gathering at the onset of a supplier problem. Sheets of this type have been found to be useful at the onset of discussions when a problem surfaces. The customer could provide the supplier with the electronic filing form, form number, severity, time reference, and the participants as well as any specific information that could be related to the flaw. The supplier can be required to fill in all relevant blanks and return the form by a designated time. This allows both the supplier and the customer to define the problem at their first meeting.

Any other provision that the customer has found to be important, depending on their industry, can be included in the requirements. Although the sheets are useful, they can be augmented with other information relevant to specific operations or applications.

Problem:				Date:	
Attendees:				Rev:	
State the problem? (Be very specific)					
State the failure mode?					
- Appearance					
- Flaw					
- Incident					
Make photographs available!					
Part containment - in plant					
- at Supplier					
- Warehouse					
- Dock					
- in transit					
Is line protected?					
- Number on hand					
- Number in transit					
- Next delivery					
- Is material dept. tied in?					
- Number to be sorted					
- When will sort end?					
- How to identify break					
- Paint marks required?					
- Where to mark parts					
- Area clearly understood					
- Mark seen after assy?					
- What color/by whom?					
# Failed / # Checked P chart					
Other plants notified?					
Make sketch with orientation and all characteristics identified					
Identification: Date code used?					
- How do you read it?					
- Number made to date					
- How many failed to date					
- How many at supplier					
Can defect be seen without disassembly?					
Unusual scratches or marks?					
All parts from same supplier?					
- produced on same Line?					
- Different patterns/cavities?					
- produced on same machine?					
- Tier2 supplier involved?					
- Process changed recently?					
Type of variable gauging used					
Downtime consistent?					
Machine rebuilt recently?					
Downtime records available?					

Figure 6.3 Table of evaluation, sheet 1.

Problem: _____

Attendees: _____

Describe process flow   _____
                   _____
                   _____
                   _____
                   _____

Process flow diagram available?   _____

Hard data and evaluation required?   _____
 - Are sketches required?   _____
 - Measure at least 10 samples   _____
  - Everything in specification?   _____
  - Same as PPAP parts?   _____
 - Compare five good and five bad parts   _____
 - Concentration diagram provided?   _____
 - Are work instructions posted?   _____
 - Are procedures availabe at floor?   _____

                     Between
                     Here and

List differences: between good and   There......
bad........   ...

_____   _____
_____   _____

Any contradictions present?   _____
                     _____

5 Why analysis results?   _____
                     _____

What is the gap in the process / method?   _____
                     _____

                     _____

Where is the biggest family of variation?
 - Time to time   _____
 - Piece to piece   _____
 - Within piece   _____
 - Other   _____

Operator input: _____

Manual supervisor input   _____

Quality assurance supervisor input   _____

Repair people input   _____

Maintenance input   _____

Other:

Figure 6.4 Table of evaluation, sheet 2.

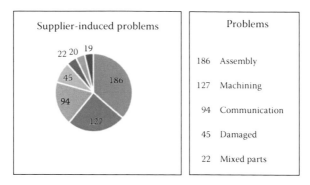

Figure 6.5 Supplier-induced problems.

Consider using these forms or having comparable reproductions of them available for your next supplier problem to expedite solutions. When you begin forming a solution roadmap it helps to ensure that the information is saved and available.

There have been innumerable investigations conducted in past problem-solving meetings with suppliers where excessive time is wasted going over the same information on a repetitive basis. Utilizing this type of sheet will provide focus and clarity to establish the basic pertinent information that is available. Once a serious problem is encountered, the customer should include any information pertinent to the supplier and should fax or electronically transfer the information, sketches, or photographs to the supplier for immediate return with their input by the next meeting.

If you can accomplish this, you will have saved more than a few weeks of meeting time and untold delays in the solution of the appearance, flaw, or incident characteristic of your problem. Follow the road map by collecting the data while developing your supplier for improvements.

Supplier visits

Supplier visits are another tool that helps to make improvements in supplier enhancement. These visits can be conducted on the supplier's premises after a request is issued by the customer. Visits for evaluation can be made prior to the establishment of a contract to supply parts as well as to conduct audits to ascertain the conditions present when a problem is created.

Generally, these visits can be used to rank the observed conditions in a relatively short period of time. Most of the visits require a few hours and can be accomplished in one day, including travel time, for local suppliers. The visit consists of a facility tour of those operations involved in providing the product or products in question. Or they can be used for follow-up to determine if agreed improvements to the PFMEA, control plan, procedures, or corrective actions are firmly in place after a problem is reported to be eliminated. Or the visits can be used as a prequalifying check to ascertain whether a proposed supplier has the systems in place to be a competent provider. These inspections are conducted to observe the important aspects of the business and will be seriously addressed in a later chapter.

Some of the key items to be observed when there is a problem under scrutiny are the DFMEA, PFMEA, control plans, work instructions, procedures, and the corrective actions that are being employed at the current time.

Supplier symposiums

The activity of conducting a supplier symposium is associated with holding a training seminar for problematic suppliers. Some suppliers are not fully experienced in the application of corrective actions to troublesome conditions. Some may not have the relative expertise to define a problem so that it can be studied. Others may lack the expertise to conduct viable studies to generate a solution. A supplier symposium helps to focus their attention on the problems that they may have caused, on the means to help them correct the problem, on the focus of problem-solving investigations, and on the certification to prove that the corrective actions taken are effective.

Certain manufacturing facilities track and record the type, significance, and frequency of problems caused by their suppliers. While working with over 200 suppliers that provided services or products to an automobile engine plant, it was not unusual to be reticent about the most troublesome 10%. This group of providers was invited to attend a meeting at the automotive plant, where a seminar was conducted for their, and the customer's, benefit.

Each of the suppliers was assigned to one of the customer quality engineers or managers, who acted as their mentor. The suppliers provided a representative or two from their facility, including plant managers, quality managers, and acting quality engineers.

Problems that suppliers cause

Each of the suppliers was welcomed and an overview was conducted providing the types of difficulties that were being experienced from all suppliers in general (Figure 6.5).

Mentor review

An overview chart can be provided for the entire supplier group or it can be tailored to focus on the independent supplier. Each type of chart permits a review of the frequency and potential impact on the customer to be discussed. Once the overview is conducted the individual mentors have the task of reviewing the problems specific to the supplier to which they have been assigned. While it is not necessary to conduct a group meeting for the symposium, it is possible to conduct this corrective operation on a one-to-one basis if necessary. Sometimes providers prefer to attend a private meeting to prevent embarrassment and to elicit information.

The mentors review the DFMEA and PFMEA revisions, as they will differ from the original documents that were awarded during precontract approval. They observe the problems at hand in view of the controls and the corrective actions that the supplier has proposed or introduced into his documents to prevent a situation occurring. The participants also discuss the severity, frequency, and occurrence ratings that have been developed for each of the troublesome operations. Detrimental conditions are highlighted and plans are assigned to establish a corrective action and the date by which it is to be accomplished. It may also be helpful in identifying a supplier weakness in the problem-solving activity. Follow-up contacts are to be enabled to ascertain that the correction timetables are being kept to.

One other item of the utmost importance should be discussed with each group of suppliers. In the discussion, it should be noted that numerous quality and production problems were experienced when a supplier had revoked and violated an agreement reached prior to the granting and obtaining of the product and process approval. It should be noted that any change in supplier tier base, materials, process, tools, chemistry, methods, or inspections without preapproved customer authorization would result in a discontinuation of that supplier's service. Compliance to this directive is necessary to ensure quality and quantity requirements. It is imperative that suppliers be aware of this requirement.

These same considerations are applicable to requests for changes to the product flow path, control plans, procedures, and work instructions. These considerations can aid supplier enhancement because they not only focus attention on the problem but encourage some corrective action to be taken. It is also possible at this time to instruct or inform the supplier representatives of the process of applying the three-point employee responsibility system of *Do Check, Don't Create,* and *Do Hold*. In addition, a couple of problem-solving aids can be introduced to help those providers that require it to acquire additional manufacturing problem identifying skills, as will be presented in a later chapter.

Problem resolution

A problem cannot be considered corrected unless it can be turned on and off at will. But because this type of verification can be too expensive and too untimely to perform, it is acceptable to use a table of comparison to determine an acceptable corrective action. This type of verification will be addressed in a later chapter that discusses potential problem-solving methods that inexperienced or new suppliers may find acceptable. This tool has been found to be extremely effective in aiding suppliers that have not received extensive problem-solving training. Its effectiveness is enhanced because the method does not require any statistical ability or calculation. Rather, it is based on an observation of test ranking or value separation arranged in tabular form.

Summary

This chapter suggested a new concept for the inspection of incoming materials. This method is enabled by the application of the employee responsibilities to peruse (Do Check), ensure work (Don't Create), and contain (Do Hold) defectives so that the products that are were using are acceptable. It referenced a series of problem resolution sheets that can be used to investigate problems when they arise. The chapter also identified a series of different meeting measures that can be used to improve supplier excellence.

These measures include five activities that should be considered when approaching supplier-related problems.

1. Immediate contact
2. Conducting problem-solving meetings
3. Visits to the suppliers
4. Supplier symposiums
5. Problem resolution

A chapter on record keeping is next. It is provided to allow attention to be directed to supplier errors and their resolutions. The chapter contains explanations of the importance of defining the problem, describing failure modes, the need for product containment, date code interpretation, and use of process flow plans. All of these are critical to creating and maintaining correct records.

chapter seven

Record keeping

The previous chapter indicated actions that can aid and correct an errant supplier. It encouraged employee involvement in the perusal, formation, and containment of the products that they provide. Information included a list of four different problem-solving communication methods depending on the significance, frequency, and occurrence of individual problems. The chapter qualified a directive for eliminating any unauthorized changes to the process irrespective of the descriptive classification that it may contain. It enlisted the requirement that no changes be allowed without prior customer authorization.

This chapter on record keeping should not be considered mundane. Recording and keeping supplier problem information will result in a system that recognizes problems and their required actions for improvement. These actions will become a valuable file of lessons learned that will be useful in the evaluation of future problems. This chapter suggests methods that will help to focus on the supplier's errors, and potential resolutions and corrections should they reoccur.

Mandate origination

The mandate "No changes allowed without prior customer authorization" was developed because of a series of individual changes that suppliers had taken to improve their bottom line. One supplier changed from using a red iron oxide to a black iron oxide with different properties that affected a sintering operation. Others changed their Tier 1 or Tier 2 supplier base to reduce manpower costs on a product that was acquired from them. Another changed the approved heat treatment facility and methods without approval and provided inadequately hardened components. There are many other examples of this, and each proved to be a costly problem for both suppliers and customers.

Now, understand that not all suppliers will make what they consider to be viable changes without approval. But there are some that do. It is worth repeating that once a process or product approval is granted at the time of contract finalization, unexpected changes are not allowed without authorization.

So, at the time of supplier contract approval there should be an approved PFMEA and control plan that include the process flow and methods of use and instruction that will be applied. This is basic to what should be required. If a supplier does not have these tools available in practice, then it is an indication that they are not fully qualified or certified. They may be willing to provide the customer with a product, but they will require intense scrutiny. Therefore, a critical consideration for the record is whether or not the supplier is certified to a national or worldwide qualifying organization (e.g., ISO, QS, CFR).

If you find the supplier to be certified, they will certainly be easier to manage as their functions will be directed to maintaining the standards that they employ. If they are not certified, it befits you the customer to guide them in the use of DFMEA, PFMEA, and control plans through a certification process. You may be able to accomplish this transition by simply insisting that their certification by an international standard is required to be your supplier.

So, these are two of the foremost reasons why record keeping is significant. One is the generation of a file that captures all lessons that were learned from a problem. The other is that if there is an approved plan with stated conditions, then it can be modified to provide improvements to prevent or reduce future problematic occurrences.

In a previous chapter, problem resolution worksheets were presented for your consideration. It is important to consider the most important features for collecting this type of data during the supplier's problem-solving process.

Statement of the problem

It is imperative that a clear and concise definition of the problem be developed. When talking to others on a phone or teleconferencing network it is not unusual that the participants may have different pictures of the condition in their mind's eye. This problem is more easily rectified when photos, sketches, and verbal descriptions are all applied. It is necessary for the participants to agree on what the problem statement defines. It is not unusual for a telephone interaction to become focused away from the problem being experienced. Someone must remind the participants of the focus that is specified in the statement of the problem. In many cases the supplier must be guided to improve their expertise in this area to allow them to become better suppliers.

Description of failure modes

It may help the supplier to understand that the type of failure experienced is of a particular classification. It could be due to the appearance, a flaw, or an incident. For example, in a molded plastic high chair tray, errors in appearance might be caused by the presence of small black flecks of regrind material. This can cause not only a visual defect but can result in a flaw whereby the boss mounting screw bases fail due to stress—hence, the importance of providing photographs or sketches with a description of

(a)   (b)

Figure 7.1 Location of fatigued high chair table mounting base: (a) acceptable mounting bosses, (b) fatigued mounting boss.

the problem (Figure 7.1). It is not an easy task to describe the defect shown in the figure to someone who does not have a simple sketch available.

The third type of problem is due to an incident. These types of problems can occur when a product is subject to severe environmental stresses such as heat, vibration, or outside force. Picture two gold wires that are employed in an electrical controller circuit (Figure 7.2). The final assembly is injected with a plasticized material that hardens to prevent lead movement or disconnection. When the flow of the hardening material becomes erratic it can move the gold wire leads significantly. If the wires are installed properly there is no chance of contact. But if the wires are installed incorrectly, one or both of the leads can move and make contact and cause a short.

I believe that you can recognize the importance of providing and requiring sketches or photographs when working with a supplier. I have included the illustrations in these chapters to convince you of their application to improving clarity and understanding. The text provided before Figure 7.2 may not be entirely clear to the reader. However, once the sketch is provided it adds more clarity to the description of the defect. Nurture your suppliers to use photographs and sketches to improve their operations and communication.

Product containment

If a supplier creates a problem within your organization it is right to assess the potential damage that it can create. This is especially true if it may impact your service to your customer. To do this it is sometimes necessary to evaluate the amount of similar materials that you have available both

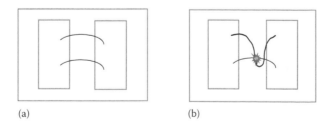

(a)   (b)

Figure 7.2 Electrical incident defect: (a) good connection, (b) electrical malfunction.

at your facility and at the supplier's. A count of suspect materials at the plant, supplier, dock, and warehouse and in transit should be conducted. Once the material has been sequestered, if necessary it can be reprocessed or disposed. It is important to establish this record to ascertain that all questionable material has been handled properly.

Date code interpretation and use

As soon as a problem is recognized the manufacturing personnel should capture a sample of the troublesome materials. In most cases these materials will contain a supplier identification number, which can be used to trace the materials to their date of manufacture. Some manufacturers did not have a date code procedure established but did have a means of identifying individual materials in a series of patterns from a molding operation. Six identical pattern inserts were each provided an identification number that could pinpoint the location of a plastic cap that was to be produced. This relevant information was beneficial in conducting problem evaluations and solutions. Encourage your suppliers to provide date codes or pattern serial numbers to allow recognition of the means of production. If this is not possible, as in the case of fluids, have the supplier certify via laboratory results that the product is acceptable for use. In any event, this information should be acquired from the supplier when applicable and should be recorded as a strong clue for problem origination.

Process flow plan

The flow plan is required to provide relevant information about the process in question. The supplier must be cognizant of the fact that any deviation from the approved PFMEA, which is a rendering of the flow plan, is intolerable. Any recognized deviation from the original plan must be corrected as expediently as possible. In many cases, problems arise when there is a divergent condition created. These different renditions of flow plans may be caused by first-piece inspections, routine inspections, rework, dropped parts, or other happenings. It is important for your supplier to recognize that parts removed from the flow plan are to be reintroduced back into the flow plan from where they were removed if they are of acceptable quality. If the supplier does not control these circumstances, problems will arise such as unprocessed final products, missed operations, foreign material inclusions, and other defects.

It might also be useful to record the following information to gain significant clues, which should be available to the customer and the supplier (Figure 7.3).

In addition to information on containment and certification, a listing of the process parameters (Figure 7.4) will be of benefit to facilitate

	Primary supplier:
	Supplier to supplier:
	Are parts contained at line?
	Are dock and shipments contained?
	Are supplier and warehouse contained?
	Is the customer line protected?
	Has the Material Dept. Been notified?
	Do others supply this part to you?
	Does your supplier have it contained?
	Have other plants been notified?
	Date code and how to read it:
	Number made with that date code:
	Number on hand and/or in stock?
	Number currently in shipment?
	How will parts be certified?
	Are certification paint/marks required?
	Where can parts be marked after sort?
	Is this area clearly understood?
	Can the mark be seen after assembly?
	What color is to be used by whom?
	How will sort containers be identified?
	How to identify sort / certification break?
	When will certified parts arrive?
	When will certification be completed?
	Did your supplier provide this info for you?

Figure 7.3 Part containment and certification list.

	Only one flow path?
	Operation done at one station?
	Are parts made on one line?
	Are there multiple flow paths?
	Machining done on one machine?
	All machining done at one plant?
	Variable or attribute gauge used?
	All parts tested on one machine?
	Is there more than one serial or pattern?
	Do all serials have the same problem?
	Do other plants use the same parts?
	Other plants % defective the same?
	Does part meet blueprint specs?
	Material changed last three months?
	Process changed last three months?
	Major equipment repair last three months?
	Variables the same as three months ago?
	Could this be an off-line repair?
	5 good and 5 bad pieces captured?
	Have good and bad been measured?
	What else can be done *now*?
	Other?

Notes:

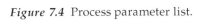

Figure 7.4 Process parameter list.

analysis, as applicable. All of the items may not be present or used in a manufacturing operation. However, a consideration of each of the areas of concern that are dependent or involved can be viewed independently for their participation.

Not all of the proposed records will be used to aid the customer in each problem. However, they are included because they may become applicable, depending on the circumstances. The important consideration is that your suppliers are developing strategies that will result in the recognition of problem areas that can be addressed and corrected. Problems that are not highlighted will not be addressed.

Summary

This chapter presented some of the reasons that record keeping is an important part of dealing with your suppliers. Establishing records creates a firm basis for understanding systems that employ the concept "No changes allowed without prior customer authorization." It also captures much-needed information that may be overlooked in problem evaluation.

Make your supplier aware of the importance of

- Problem statements
- Failure mode descriptions
- Date code interpretation and use
- Process flow plans
- Product containment

Accordingly, the information recorded becomes very important as it enables you to recognize problems that require actions and solutions. These improvements may require revisions to the DFMEA, PFMEA, and the control plan, which will aid you to enrich the supplier.

The next chapter deals with those considerations that are to be evaluated when a supplier is being considered for a new activity or contract. The information can also be used to qualify a new supplier for approval. In addition, the observations can be used to evaluate why a supplier may be supplying an unsatisfactory product. The chapter will provide a start for those that have no formal system and will provide insight into areas that have not been formally considered.

chapter eight

Supplier readiness

The previous chapter stressed that accurate record keeping is essential when gauging supplier performance and readiness. The records generated will aid in creating a firm basis for understanding current problems and may indicate potential actions for problems experienced in the future. Accurate statements of the problem, descriptions of the failure mode, date code use and interpretation, process flow plans, and the necessity of having product containment plans are obligatory considerations. Process and supplier improvements are created by rectifying adverse conditions and addressing them in the design, the process, and the process control plan.

This chapter deals with those concerns that are to be evaluated when a supplier is being considered for a new activity or contract, or the information can be used to gauge a new supplier's qualifications for approval. In addition, the observations can be used to evaluate why a supplier may be supplying an unsatisfactory product. The chapter provides a start for those that have no formal system and can provide insight into areas that have not been formally considered.

When conducting this type of inquiry it is necessary to list the identifying information that describes the project at hand. For instance, see the listing in Figure 8.1.

A customer may find other information useful when conducting an assessment. However, the list presented will provide a good basis to start evaluating suppliers. Once the information is provided it is then necessary to assess the specific quality and delivery capabilities of the supplier under consideration.

The information pertinent to assessing the supplier lies in understanding their capability in at least four areas of interest. The four areas of interest are affected by the need for some suppliers to be included in the design phase of a project. These four areas, as applicable, are as follows:

- Is developmental and technical information available? If the supplier is involved in the design of the part, do they have the capability of using DFMEA and PFMEA tools and requirements?
  - Are the latest drawings available?
  - Are all tolerances finalized?

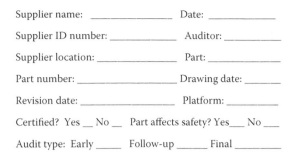

Supplier name: _____ Date: _____

Supplier ID number: _____ Auditor: _____

Supplier location: _____ Part: _____

Part number: _____ Drawing date: _____

Revision date: _____ Platform: _____

Certified? Yes __ No __ Part affects safety? Yes___ No ___

Audit type: Early _____ Follow-up _____ Final _____

Figure 8.1 Supplier assessment list heading.

- Are the final approved drawings on hand?
- Has DFMEA been developed?
- Has PFMEA been developed?
- Is the control plan completed?
- Does the supplier have a documented quality plan?
  - Is the PFMEA acceptable?
  - Is this a previously established PFMEA?
  - Are there adjustments listed in the PFMEA?
  - Is a flow path instrument available?
  - Do the flow path and PFMEA match?
  - Are critical PFMEA areas addressed?
  - Are inspection types and frequencies acceptable?
- Can the quality plan be executed?
  - Do work stations match the flow path?
  - Are operator instructions posted and available?
  - Can operators explain the instructions available?
  - Are the necessary tools available on the job site?
  - Is the lighting adequate?
  - Are all measuring instruments and gauges properly calibrated and identified?
  - Are all incoming materials given a visual inspection before use?
  - Can operators describe the actions to be taken if something is amiss?
  - Is there a designated holding or do hold area for nonconforming items?
  - Are containers and flats checked before any product is loaded into them?
  - Are all containers, parts, and inspection items properly tagged for identification?

- Do all inspections and checks follow the sequence and method specified?
- If statistical process control charts are displayed, do they make sense in terms of stratification, runs, trends, or other anomalies?
- Are there reasonable control limits displayed, and do they show adjustments for causal conditions?
- Is the process capability (Cpk) established?
- Is the enactment of quality systems evident?
  - Are work instructions posted?
  - Are specified tools available on the job site?
  - Are scrap and rework charts available?
  - Is a problem resolution method in practice?
  - Is there a problem resolution display area?
  - Is information passed shift to shift?
  - Is information passed area to area?
  - Do operators get timely information?
  - Do operators feel involved in problem solving?
  - Are operator interests addressed?
  - Is there a system to prevent mixed parts?
  - Is there evidence of PFMEA and control plan revisions for the last problem encountered?

These are some of the line items that should be addressed when evaluating the acceptability of a new supplier. They can also be used as a performance check system to evaluate the current status of suppliers currently involved in supplying materials who have contributed to recent problems.

Other supplier features

Depending on the circumstances, it may be necessary to track other features that are systemic to the supplier being evaluated. These may include the capabilities of those that supply them. It may also include a timetable to show the installation time required for the process equipment. Future equipment must be included in the flow plan and be included in the relevant PFMEA and control plan documents. In addition to the equipment requirements, there should be a recognition of the manpower requirements, both skilled and unskilled, to obtain the desired manufacturing levels. Manufacturing and skilled employee requirements should be determined by local availability. Is there an effective maintenance schedule that can be documented? Is there a preventive maintenance program in effect? Improvements in any of the areas will enable an increase in supplier performance.

Supply chain concerns

If the supplier has a Tier 1 or Tier 2 supplier, do they have the systems in place to ensure acceptable performance? If not, then the supplier being evaluated must either develop a capable source or secure another qualified supplier. Future equipment must be included in the flow plan and be included in the relevant PFMEA and control plan documents.

Installation concerns

Does the supplier chain have the capability and capacity to meet the desired manufacturing schedule? If future equipment must be installed, what are the projected schedule completion dates? If this equipment is promised by local or foreign manufacturers, will all training be completed in order to meet the production schedule? Does the scheduled equipment run-out date provide enough reaction time in the event that problems occur? Once the equipment is installed and run-out, the operation requirements must be included in the flow plan, PFMEA, and the control plan. In a later chapter, information will be provided to help the supplier address manufacturing problems so as to improve operations. Some suppliers do not have the manufacturing capability and can be focused on preventing costly scrap or scheduling maintenance to prevent rework.

Manpower concerns

Depending on the supplier location, it befits the supplier to attain acceptable levels of staffing to fulfill their commitments. Generally, in the current marketplace it is possible to obtain relatively unskilled employees who can be educated and trained to perform the more ordinary tasks. This also appears to be true for more educated college graduates who are available and proficient in their areas of expertise. Lately, most staffing difficulties come in the form of filling skilled trade positions, which require knowledgeable individuals to perform difficult mechanical tasks safely. These positions include construction, machining, pipe fitting, and electrical, sheet metal, and machine repair operations. If your supplier does not possess adequate skilled and unskilled employees to perform the required tasks, it will impact their service quality and timeliness.

Maintenance concerns

In addition to the manpower of skilled maintenance staff is the requirement to possess an effective maintenance schedule. Basic to all the schedules is the use of lubrication to oil and grease the operating transfer and forming machines (transfers, rams, lathes), transportation devices (locomotives, cranes, belts, lifts, machine ways), and individual components (air bottles, cylinders, hydraulic tanks). Notwithstanding, other important criteria should be considered for change and repair if the supplier does not act on an as-needed basis. Certain actions, such as furnace lining and refractory replacement, oil changes on vehicles and machines, and belt or bearing changes on moving equipment, will prevent catastrophic failures and will provide a smoother manufacturing operation. The presence of a preventive maintenance program adds a layer of confidence to the belief that the supplier is competent.

Major issues and concerns

Is the supplier experiencing any major issues at the current time? Do they anticipate any major issue in the future? What is their backup plan in the event of a natural catastrophe? These are all major concerns that the customer needs to consider before creating a supply contract. If the supplier has not yet considered these issues, it is timely that they do so to help to protect their customers.

Something as bland as supplier gifts can cause problems that have not been anticipated. Although this may appear to be insignificant, suppliers should be assessed to ensure that they do not provide gifts upon visiting your facility and that they are qualified to provide the services requested. The reasons for this will become self-apparent as we consider the following three cases of concern.

Case 1: Wooden rulers

A supplier visited a machining operation and wanted to make a good impression on the machinists. He passed out wooden rulers with his company's name impressed and printed on one side. Unfortunately, at the next QS9000 and ISO audit the wooden rulers were observed in the plant area without a certification sticker and without being recorded in the gauge system. Necessarily, this presence resulted in an unacceptable finding during the audit evaluation.

Case 2: Whiteout

A customer was visited by a well-intentioned supplier, who provided small bottles of whiteout, which is a correction fluid that can be used to correct typing or writing errors. This unauthorized material was not accompanied with a material safety data sheet (MSDS), which was required to be included in the customer's file. This also resulted in an unacceptable finding during an audit that affected the supply chain to their customer.

Case 3: Unauthorized sign

A plant had a serious accident because an employee failed to wear a safety harness while performing overhead work. A visit from the Illinois Environmental Protection Agency (IEPA) was conducted over the entire plant property. The IEPA conducted a wall-to-wall inspection of the plant on all levels. The plant was cited only once for one violation of federal requirements.

The citation was for having an 18-by-24-inch sign on a transformer doorway that was too large according to federal requirements, resulting in a $10,000 fine. An authorized supplier had supplied and installed the sign in a previous period but had overlooked the existing government regulations. The customer plant did not specify the mandated size of the replacement signs to the supplier.

As it was, the plant appealed the citation on the grounds that the intent of the government regulations had been met. The sign warned of the presence of high-voltage conditions within the locked enclosure. It also specified that there were hazardous liquids contained within the transformer bodies. The complaint was withdrawn only after much difficulty. This was an unnecessary investment in time to have the sign accepted.

No one needs problems like these. This chapter contains a lot of questions that may appear to be too troublesome to ask. In addition, there are some conditions that may not be prevented by a supplier improvement plan. However, it is better to be prepared and to take preventive actions rather than to react to extraneous conditions that might have been stopped before they occurred. The heading on a supplier assessment list should contain the pertinent information that will provide explicit identification of a specific case or problem (Figure 8.2).

The previous sections show how simple it is to originate a supplier assessment plan. It should be the starting point for reviewing all proposed or existing suppliers to help prevent future problems. Since it is important, much of the material has been repeated because of its significance to stimulating improvements.

Summary

This chapter dealt with supplier readiness—those considerations that are to be evaluated when a potential or current supplier is being considered for a new activity or contract. The information can be used to qualify a new supplier for approval. In addition, the observations can be used to evaluate why a supplier may be supplying an unsatisfactory product. The chapter contains provisional supplier assessment sheets that can be used as a starting point for your enterprise. If no formal system is currently

Supplier name: _____   Date: _____

Supplier ID number: _____   Auditor: _____

Supplier location: _____   Part: _____

Part number: _____ Drawing date: _____

Revision date: _____   Platform: _____

Certified? Yes __ No __   Part affects safety? Yes___ No ___

Audit type: Early _____   Follow-up _____ Final _____

- <u>Is supplier planning evident?</u>   Yes___ No _____

  1. Are the latest drawings available?   Yes___No_____

  2. Are all tolerances finalized?   Yes___ No _____

  3. Are the final approved drawings on hand?   Yes___ No _____

  4. Has the DFMEA been developed?   Yes___ No _____

  5. Has the PFMEA been developed?   Yes___ No _____

  6. Has the control plan been completed?   Yes___ No _____

- <u>Dose the supplier has a documented quality plan</u>   Yes___ No _____

  1. Is the PFMEA acceptable?   Yes___ No _____

  2. Is this a previously accepted PFMEA?   Yes___ No _____

  3. Are the adjusments listed in the PFMEA?   Yes___ No _____

  4. Is a flow path instrument available?   Yes___ No _____

  5. Do the flow path and PFMEA match?   Yes___ No _____

  6. Are critical PFMEA areas addressed?   Yes___ No _____

  7. Are the inspections and frequencies acceptable? Yes___ No _____

- <u>Can the quality plan be executed?</u>   Yes___ No ___

  1. Are instructions posted and available?   Yes___ No ___

  2. Can operators explain the instructions?   Yes___ No ___

  3. Do workstations match the flow path?   Yes___ No ___

Figure 8.2 Supplier assessment list.

Supplier name: _____ Date: _____

Supplier ID number: _____ Auditor: _____

Supplier location: _____ Part: _____

Part number: _____ Drawing date: _____

Revision date: _____ Platform: _____

Certified? Yes __ No __   Part affects safety? Yes___ No ___

Audit type: Early _____   Follow-up _____ Final _____

1. Are the necessary tools available on the job site?   Yes___ No ___

2. Is the lighting adequate?   Yes___ No ___

3. Are measuring devices and gauges calibrated?   Yes___ No ___

4. Have the entering materials had a visual

   inspection before use?   Yes___ No ___

5. Can operators describe the corrective actions

   to be taken if something is amiss?   Yes___ No ___

6. Is there a designated hold or sequester

   area for nonconforming items?   Yes___ No ___

7. Are containers and flats checked before

   any product is loaded into them?   Yes___ No ___

8. Are all containers, parts, and inspection

   items properly tagged for identification?   Yes___ No ___

9. Do all inspections and checks follow

   the sequence and method specified?   Yes___ No ___

10. If statistical process control charts are used,

    do they make sense as far as stratification,

    runs, and trends?   Yes ___No___

11. Do the charts show reasonable control limits?   Yes___ No ___

Figure 8.2 (Continued) Supplier assessment list.

Supplier name: _____ Date: _____

Supplier ID number: _____ Auditor: _____

Supplier location: _____ Part: _____

Part number: _____ Drawing date: _____

Revision date: _____ Platform: _____

Certified? Yes __ No __ Part affects safety? Yes___ No ___

Audit type: Early _____ Follow-up _____ Final _____

1. Are the control charts trending or out of control? Yes___ No ___

2. Is process capability (Cpk) established? Yes___ No ___

3. Is the enactment of quality systems evident? Yes___ No ___

4. Are the work instructions posted? Yes___ No ___

5. Are the specified tools available on job site? Yes___ No ___

6. Are scrap and rework charts available? Yes___ No ___

7. Is a problem resolution method in practice? Yes___ No ___

8. Is a problem resolution display area present? Yes___ No ___

9. Is information passed from shift to shift? Yes___ No ___

10. Is information passed from area to area? Yes___ No ___

11. Do operators receive timely information? Yes___ No ___

12. Are operators involved in problem solving? Yes___ No ___

13. Are operator interests addressed? Yes___ No ___

14. Is there a system to prevent mixed parts? Yes___ No ___

15. Are error-proofing measures used? Yes___ No ___

16. Is a strong containment plan in use? Yes___ No ___

APPROVED: NOT APPROVED: REAUDIT DATE:

Yes___ No ___ Yes___ No ___ _____

Reason for nonapproval: _____

Figure 8.2 (Continued) Supplier assessment list.

available, the materials and thoughts contained within the chapter will enable supplier evaluation and result in supplier improvement.

Of course, a customer must be interested in improving their suppliers, which requires effort. But be aware that there are vast benefits available when the efforts described are applied to suppliers. Better service, reduced costs, improved profits, and supplier improvements will be obtained after suppliers are mentored in these conditions.

The next chapter contains reflections on how to address a frequently experienced supplier problem.

chapter nine

Opportunities for improvement

The previous chapter looked at supplier readiness. It dealt with those concerns that are to be evaluated when a potential or current supplier is being considered for a new activity or contract, or the information can be used to qualify a new supplier for approval. In addition, the observations can be used to evaluate why a supplier may be supplying an unsatisfactory product. The chapter contains provisional supplier assessment sheets that can be used as a starting point for your enterprise. If no formal system is currently available, the materials and thoughts within the chapter will enable supplier evaluation and result in supplier improvement.

This chapter discusses the necessary steps to provide less experienced suppliers with the expertise to improve their systems. If they are unaware of these causative factors they will not recognize the potential for failure. And, in addition, these suppliers may not be aware of the significance of problems caused by mixed parts in the customer application. The problem conditions to be discussed are related to the frequency of receiving mixed parts in various industrial applications. Mixed parts can cause line jams, improper assembly, and related safety problems. Pinpointing potential problem-causing conditions will lead to improvements in both the system and the quality of product that the customer receives.

Improvement opportunities

This section will focus on one major shortcoming that can cause a potential problem in almost any industry. It involves receiving a contaminated shipment of parts in which some do not meet the specifications because of their dissimilarity. Basically, two individual parts of different design or specification in a shipment are *mixed*. The best way to eliminate this type of problem is not through a sorting process. Rather, this condition can be prohibited by preventing the parts from being mixed in the first place.

Prevention

Prevention is the keyword in ensuring that dissimilar parts are not received in the same batch or shipment. Prevention is the observation and control of detrimental activities or conditions that have been recognized as affecting a system or causing a problem. Prevention activities are associated with

- Container use and practice
- Sorting concerns
- People concerns
- Training concerns
- Assignment responsibility concerns
- Machining area responsibilities
- Station part control concerns
- Mixed-part equipment concerns
- Methods used
- Other conditions

All of the above circumstances can contribute to a condition involving mixed parts in numerous industries. The following sections will list those conditions that were found to be causative. They may not be applicable to all suppliers as they were developed from numerous suppliers. This is advantageous because the concerns can be applied to any supplier base where individual questions may be relevant or ignored.

Container use and practice

Now, some would not consider the use of containers to be a large problem in most industrial settings. But there are many manufacturing and assembly plants that are confounded by problems that may be caused by container misuse or disrepair. Damaged containers can cause part leakage, part hang-up, and even part loss in the handling process.

If containers are returned from a customer plant after parts have been unloaded and are not checked to ensure that old tags and parts have been removed, they can create an environment that leads to mixed or misidentified parts. These conditions provide opportunities for error for which the supplier will eventually be blamed. And conversely, if the supplier does not check for damaged containers and retained parts then there is a definite risk they will provide mixed parts to their customers. It befits both the supplier and the customer to be aware of conditions that can affect their operations.

The abbreviated schematic in Figure 9.1 can be used as a starting point for suppliers that have no established system in place, if it is applicable to their industry. If your supplier occasionally provides shipments with mixed parts, it would do them well to consider the incorporation of a flow plan to help prevent mixed-part conditions.

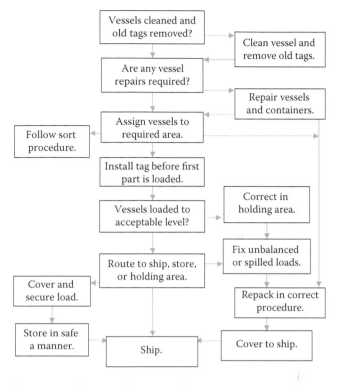

Figure 9.1 Container plan to reduce mixed parts.

Any system will provide better customer protection than no system at all.

The following considerations are applicable to containers and their use. Containers are any vessels used to store parts (Figure 9.2).

This list contains a colossal amount of information that seems to be too difficult to handle effectively. This is not a correct impression. Once a method has been established that specifies the harmful conditions, they can be easily checked by conducting a simple audit within a few minutes while passing through different departments of a supplier. This type of observation is especially applicable to a supplier that has been providing too many mixed parts.

Sorting concerns

The sorting operation is another area of concern where lack of procedures and acceptable methods can result in the loading of mixed parts. This type of activity may be experienced at the beginning or end of a heat-treating

- A clean container procedure has been established.

- A container flow chart is available for review.

- Container problems have been defined in the DFMEA.

- Container problems are corrected in the PFMEA.

- Containers are removed for repair when they can allow part retention or hang-ups.

- Damaged or leaking containers are quarantined.

- Empty and reusable containers are not used to collect trash or scrap.

- Containers and skids are empty and clean before use.

- Open containers of components are not present.

- Stacked containers do not spill parts.

- Skids are loaded only to a safe height and content level.

- Completed loads are covered with a sheet or shrink wrap.

- Full containers in storage must be covered.

- Dunnage containers are not used for part storage.

- Manpower has been trained to comply with procedures.

- Open, uncovered shipping containers are not used to store parts.

- Operators are responsible for checking containers, skids, and flats before use.

- ID tags are placed on containers before first part is loaded.

- Use only new cardboard cartons if this type of container is used and required.

- Use only undamaged molded basins, skids, or flats where applicable.

- Processing floor area does not contain unnecessary or strewn parts.

- Suppliers must insist that their suppliers provide container control for usage and cleanliness.

Figure 9.2 List of container considerations to prevent mixed parts.

furnace, for example. It can also be experienced at the shakeout end of a molding operation where different parts are molded simultaneously. In fact, it can even occur in a generic sorting operation where many loads of spilled parts are collected so that they may be separated for inspection or shipment.

The following flow plan indicates some of the sorting parameters that should be considered. Inspections, cleaning, tagging, loading, and storage

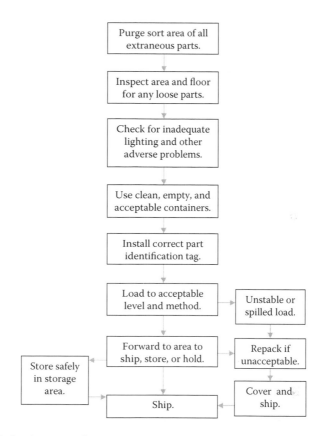

Figure 9.3 Sorting considerations to prevent mixed parts.

conditions can all enhance the supplier's ability to control mixed-part problems (Figure 9.3).

The sorting considerations shown in Figure 9.4 are related to those operations where a mix of parts is viewed for separation, inspection, or shipment. Therefore, a sorting operation is identified by an area where attempts are made to separate parts that have become mixed. The schematic should be the basis for consideration to prevent mixed parts. Again, these considerations can be used as the starting point for suppliers that are experiencing mixed-part problems.

The list identifies many different weaknesses that must be overcome if a valid sorting is to be conducted. The reasons for failures are experienced by both qualified suppliers and those lacking competent skills. Your supplier, whether internal or external to your operation, should be mentored to improve their skills. These items can also be easily verified by conducting a simple audit that can be accomplished via a simple walk through the

- A sorting procedure has been established.
- A sorting flow chart is available for review.
- Sorting problems have been defined in the DFMEA.
- Sorting problems are corrected in the PFMEA.
- Adequate lighting is available for identification.
- The floor has been painted to highlight dropped parts.
- The sort area is a segregated location.
- Lateral protection is provided to prevent sorted parts from falling to the floor.
- Close open tables under the sort table to prevent the entrapment of parts being sorted.
- The sort area is kept separate and secure.
- Purge automatic sort machines before starting.
- The area is purged of unnecessary parts before the sort is to begin.
- Pictures of parts to be sorted are provided as required.
- Position parts correctly for automatic camera sorting.
- Use shields or masks to sort parts.
- Inspect and use new empty containers for the sort.
- Sort parts directly into empty and acceptable containers.
- Immediately pick-up parts that are dropped.
- Immediately tag all nonconforming parts and move them to the nonconforming do hold area.
- Tape and seal containers in the sort area.
- Update work instructions as may be required.
- Comply with the established container check-list.
- Audit the sorting and packaging operation at irregular intervals.

Figure 9.4 Required sorting activities.

sorting area. Once the procedure is established, it can be accomplished in a few minutes.

People concerns

The following considerations are related to people concerns. The listing also identifies those operations that are necessary to prevent mixed parts from occurring. Some of the concerns are posed as statements, whereas others are stated in question form. The internal or external suppliers should be capable of informing the customer of their capabilities in those areas that are relevant to their operations.

The following check sheet identifies those that may also affect the process of mixed parts due to mishandling. All should be trained in the handling of found or spilled parts (Figure 9.5). The schematic is directed (keyed) with the use of arrows to indicate the determination parameters. In so doing, the arrows show the dedicated path of processing or those concerns that should be addressed.

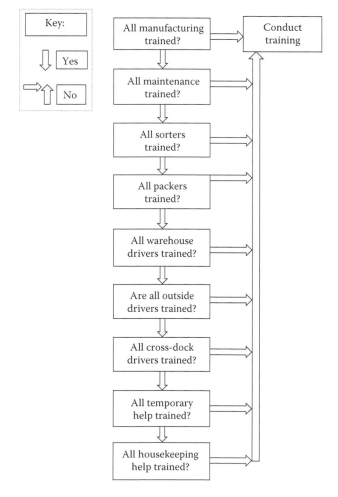

Figure 9.5 Training concerns to prevent mixed parts.

Now, in addition to having trained personnel to aid in performing part identification and traceability protection, it is also beneficial to have responsibilities assigned to mandated operations. For example, the following considerations might be present in the customer's or supplier's operations. Someone is required to perform these checks whenever a process change is enacted. These conditions are not present in all facilities, but they are commonplace in many industries and may help any supplier that is providing their customers mixed parts (Figure 9.6).

So, to briefly sum up, where people are involved it is necessary to establish a part containment procedure with a flow path that applies training and assignment methods to improve the supplier base. If the supplier does not have these in place, they may be subject to mixed-part errors.

In addition, there is a necessity to establish corrective methods where machines or machining are involved.

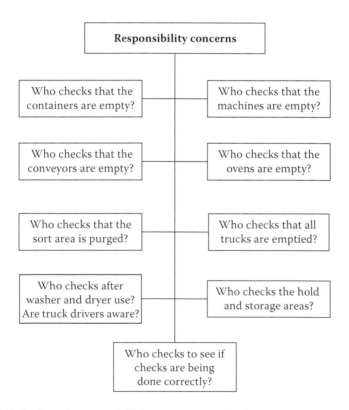

Figure 9.6 Assigned responsibilities to prevent mixed parts.

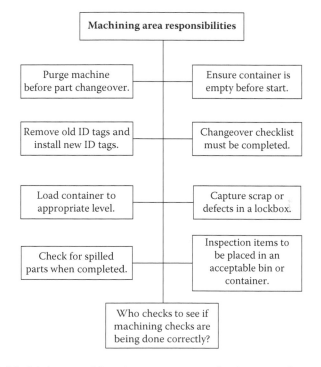

Figure 9.7 Machining considerations to prevent mixed parts.

Some of these considerations are shown in Figure 9.7.

In addition to the machining operators there are other supplier employees that can be involved in the creation of mixed-part loads. Consequently, the supplier should provide job instructions that specify procedures to prevent the occurrence of mixed parts. See Figure 9.8 for areas that should be under scrutiny for mixed-part control.

There are many other causes for mixed parts, but let me end with those equipment concerns that are the most frequent or the most troubling. These are included in Figure 9.9.

All of the items that are the focus of the figures caused mixed parts at the customer location. Now, not all of these will be present at suppliers that provide mixed parts. However, frequent violators can improve if they begin a program that will audit the troublesome conditions in their facility on a regular basis. Conducting this type of audit is not a challenging or difficult task. As I related earlier, experience shows that a walk through the facility at irregular but timely intervals will allow the observer to recognize those conditions that must be corrected. If corrective actions are to be made by problem suppliers that resist change, then it might be beneficial to demand the corrective action required.

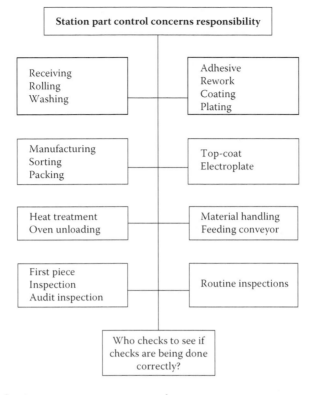

Figure 9.8 Station part component procedures.

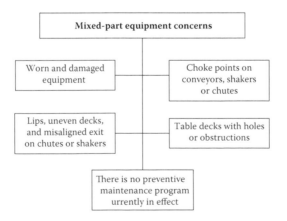

Figure 9.9 Mixed-part equipment concerns.

Each unfavorable condition observed during the audit must be addressed for correction.

Method used

All of the previous considerations are important. These next few conditions are decisive elements that must be checked for compliance. These are simply the methods that must be specified, required, and used. Included in these is compliance to the DFMEA and PFMEA requirements, including individual job instructions and work methods that are to be employed to perform the job satisfactorily. The DFMEA should consider the conditions related to the design of the system being used. The PFMEA should consider the process conditions required to ensure quality and quantity. The individual instructions and work methods are then focused on complying with the analyses to perform the work in the manner specified with the correct tools in a particular sequence and manner.

Other concerns

The control of mixed parts is not only a Tier 1 supplier concern. It also concerns internal suppliers, who provide input to other functions within a customer facility. Nothing can aid your suppliers more than obedience to mandates that have been specified to control mixed parts both within and outside the customer facility.

Summary

Of course, a customer must be interested in improving their suppliers, which requires the efforts described. But be aware that there are vast benefits available when the efforts are applied. Better service, reduced costs, capable processes, improved profits, and other supplier improvements will be obtained after mentoring them on their performance.

This chapter provided many insights into the means to identify and control the problem of mixed and sometimes damaged parts. It explains the necessary steps to providing less experienced suppliers with the expertise and methods to improve their systems. If they are unaware of these causative factors, they will not recognize the potential for failure. And, in addition, these suppliers may not be aware of the significance of the problems caused by mixed parts in the customer application. The problem conditions to be discussed are related to the frequency of receiving mixed parts in various industrial applications. Mixed parts can cause line jams, improper assembly, and related safety problems. Pinpointing potential problem-causing conditions leads to improvements in both the system and the quality of product that the customer receives from the supplier.

The biggest opportunity for supplier improvement with the mixed-part problem is the concept of prevention. Prevention of a problem is much more effective than reaction to that problem. That being said, when a new occurrence of mixed parts is identified, the customer and supplier should work together to perform corrective actions, analysis, and resolution. This problem solving can implement containment if necessary, select likely causes, change the systems, apply corrective actions, and verify the results obtained. This chapter provides a broadly defined base of information for your consideration. The listings provided can be considered for incorporation into your existing methods, as may be applicable.

The next chapter presents methods to enable suppliers to be more efficient in solving simple problems.

chapter ten

Problem resolution aids

The previous chapter dealt strictly with the condition of mixed parts, which can affect numerous types of customers. The chapter listed some opportunities for improvement that can be considered to develop all suppliers. In summary, the supplier needs to provide proper controls of plant operation so that it can have a direct positive impact on the customer's profitability and operation. The requirements are not all-inclusive, but they do represent the majority of provisions that should be practiced by all suppliers. They can be utilized to create an effective system for inexperienced suppliers.

This chapter will provide a simple method to permit the inexperienced or inept supplier to improve their problem-solving abilities. It will illustrate more fully why defective materials should not be accepted at a job site; should not be created by machining, forming, or assembly; and should not be passed on to the next operation or customer.

Three items of compliance

In previous times there were not many specific guides to aid problem solvers in the development of a plan of attack to help their employees to perform their work correctly. Over the years it has been noticed that employee involvement is required to achieve acceptable quality levels. But it has also been found to be necessary to provide employees with simple rules or methods to allow them to achieve this goal. These three rubrics of compliance were introduced in Chapter 2 and are shown in Figure 10.1. They are worth repeating here again.

"Check inputs" simply means "Do Check." "Maintain quality" means "Don't Create." "Hold all flaws" means "Do Hold."

These three restrictors can be considered to be controls and can be applied when using any comparable analysis tool when the procedural, mechanical, electrical, or hydraulic root cause is unidentified. It is an additive in that it can provide an additional level of protection or information to the process evaluation to ensure better-quality operations.

Before we focus on these defect restrictors, an example problem will be discussed to relate how these controls can aid the supplier to correct unfavorable conditions. A brief description of the process follows to enable the reader to understand the flow of the problem, the investigation

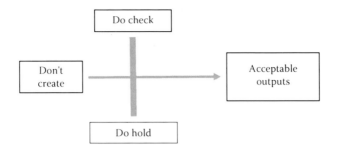

Figure 10.1 Trilogy of supplier enhancement.

conducted, the corrective actions taken, and the need to verify that the corrective actions were applicable and effective.

Problem history

A large engine plant experienced a rash of unacceptable leakage from front engine covers. The covers were leaking in an area where seals were installed to encapsulate the crankshaft and to prevent leaks. Figure 10.2 shows a front cover with a supplier seal already installed. It had the potential to leak at the inside diameter after it was installed on the engine assembly.

Seals were received from the supplier and were loaded into a tube by an hourly employee, which allowed them to be indexed for separation and automatically transferred to the location for the front cover seal assembly. The seal was installed in the front seal housing after being transferred onto a positioning fixture, where it was mounted on the engine cover. The

Figure 10.2 Front cover with seal installed.

engines were transported by conveyor through many final assembly operations before and after the seal installation. At the end of the production line the engines were tested on test fixtures to ensure that all operations had been completed correctly. It became apparent that there were excessive engines being rejected at the end-of-line test because of leaking front seals.

Everyone involved observed the installation operation, which was not able to be scrutinized effectively. The device mating the engine block on the fixture was an automatically fed ram that inserted the seal into the assembled cover. The entire area view was blocked by the engine and housing during assembly. This blocked compact area can be observed in Figure 10.3, where the pusher section remains open and unobstructed. In addition, the problem had a low chronic frequency and was accompanied with excessive reject spikes occurring irregularly during the insertion operations.

At the beginning of the study a few engines that were rejected were tested on a secondary test stand to verify that the primary test stand was operating adequately.

A chart with results from eight engines that were tested on two different test stands confirmed the results and provided acceptability and test validity for only four engines (Figure 10.4). It was believed that the problem was in the testing machines, as indicated on the chart. After correcting the testing machines, the problem persisted at a similar rate of occurrence.

Figure 10.3 Front cover pusher assembly.

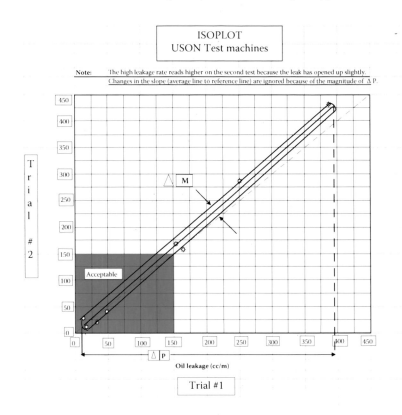

Figure 10.4 Comparable test results.

Figure 10.5 Damaged front cover seal.

Most defective seals were not readily observable for damage when in the installed position. Only after removal was the extent of the damage apparent. Some seals were undamaged, while others had significant damage, as shown in Figure 10.5.

In Figure 10.5 it can be observed that the metal ID was significantly deformed. In some cases the damage appeared to be insignificant. In other samples the outside diameter of the seal, which contained a rubber coating, appeared to be scuffed. The presence of scuffing can be seen at the very top of the seal at the 12 o'clock position, indicated by a short piece of string. The internal aberrations present on the internal diameter of the seal were caused by grease retention on the component.

In some cases it is necessary to rely on the suppliers' expertise in solving problems. This can be problematic in that some suppliers may have limited problem-solving skills. Suppliers ultimately attempt to provide the best responses to dissatisfied customers to maintain the customer–supplier relationship. At times this lack of skill becomes bothersome, and the customer is advised to mentor the supplier to solve the problem. Some important tools will be discussed in a later section to aid the customer in mentoring the supplier.

Figures 10.6a and 10.6b represent an investigation as to the cause of the leaking front cover seals.

Problem: Front seals are leaking at the ID   Date: 9/19/xx

		Condition:	Action:
?	1	Seal/crank shaft/torsional balancer part are not size.	• The mating parts have been inspected and are to specification.
	2	Rubber is damaged upon intallation.	• All leaking ID seals inpected with 30X microscope do not show any rubber damage.
	3	FM. misalignment, or damage prevent sealing.	• Some type of FM. misalignment or damage prevent ID seal from satisfactory sealing.
?	3a	Cranks half/balancer surface contains FM.	• Review of sample surfaces did not how any foreign material on mating surface.
	3b	Torsioner nose surface contains FM.	• FM and wer RP residue were present and corrected 10/1.
	3c	Seals contain FM from supplier.	• Seals sampled did not show FM from supplier.
	3d	Seals contain FM from Redicoat process.	• No new seals were observed with redicoat contamination on the ID surfaces.
	3e	Seals are contaminated during seal installation process.	• Ram well is now cleaned at least once on each shift.
	3f	Contaminated and rejected seals are reused.	• Process was changed to prevent reintroduction of dirty and damaged seals into the system.
?	3e1	Grease collects on the seal ram head and in the ram housing blind area below the ramp usher.	• Overspecification grease in now controlled.
	3e2	Redicoat material is retained on the ram shaft when the ram is retracted.	• Ram well is now cleaned at least once on each shift.
	3e3	Found supplier grease out of spec.	• Supplier adjusted process and provided capability stydy for grease.
	3e4	Some seals are stuck off-set to each other.	• Supplier adjusted precess capability and is layer auditing for compliance.
	3e5	Greases amount not to specification.	• Supplier adjusted process capability and is layer auditing for compliance.
	3e6	Seals stuck in offset position can hang up and be damaged in system.	• Supplier and tonawanda separate with light thumb Pressure.

Figure 10.6a Facing page of front seal study.

?	3e1a	Grease on seals is over the specification. • Supplier provided capability study and is auditing for compliance.
	3e2a	Redicoat ovals are retained on ram shaft or fall into blind well. • Cleaning daily prevents accumulation of grease and FM in the ram well.
	3e2b	Redicoat in unwanted areas. • Cleaning daily prevents accumulation of grease and FM in the ram well.
	3e2c	Redicoat in unwanted areas. • Cleaning daily prevents accumulation of grease and FM in the ram well.
	3e5a	Grease amount excessive. • Supplier sent sample of grease amount scoring transform to help them control amount of grease in seals

?	3e2a1	Retraction of ram head allows bushing to wipe the Redicoat into unwanted areas. • Cleaning ram well daily
	3e2b1	Material left on the ram is in the crush area between the ram shoulder and the seal, which creates FM pieces • Cleaning ram well daily.
	3e2c1	Stringer and crushed Redicoat material is trapped in the blind well below the ram and can be carried into the seal • Cleaning ram well daily.

Finding: OD leaks were eliminated with the daily cleaning of the ram well. ID leaks may be due to the installation process and FM, which has been corrected.

Action: Test shows significant differences due to the installation process.   **Visual keys** = movement and misalignment during installation.

Recommendation:

1. Supplier must regain control of their grease application process and supply seals with grease within the specification. **OK 8/29**

2. Until the seal ram housing can be reworked to allow exit of excess grease and Redicoat streamers, the unit must be cleaned under a PM schedule that involves a minimum of 1 times per day. **OK Started on 1st shift and 2nd shift on 8/13**

3. Supplier has been requested to provide updated capability and layered audit information by 8/29. **OK 8/29**

4. Eliminate movement of case during assembly, check stop pads & locators, align feeder rails and seal ram to crankshaft. **OK 9/29**

Figure 10.6b Facing page of front seal study.

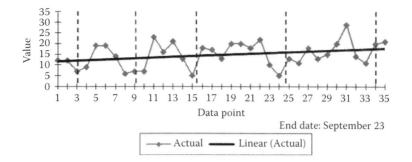

Figure 10.7 Leaking front cover seals: Actual defects, both shifts combined.

The problem was very serious; not only did the plant have to refurbish the affected engines, it also had to reprocess them through the assembly and testing system.

Note that in Figure 10.7 there is an increasing trend in the number of leaking oil seals. Understand that there were in excess of 500 repairs that had to be conducted in a little over a month. Since the line ran at a rate of four engines per minute, this resulted in a loss of 125 minutes of production line downtime. Even at a loss rate of $100 per minute this represented a monthly loss approaching $12,500.

Everything explained thus far indicates the effort that was invested in solving the leaking front seal problem. Significantly, these efforts did not lead to a solution to the problem.

Leaking front seal problem solution

One day, while an engineer was walking through the assembly system looking for detrimental conditions during an audit, he made a discovery. At the platform where the supplied seals were loaded into the automatic loader, the engineer noted many seals strewn about the remote loading platform. The loader was immediately requested to report to the site to discuss this condition. The loader indicated that the supplied seals were not consistent. Sometimes the seals appeared to be stuck together differently because they could not be loaded easily into the feeder chute. Asked what the problem was, the loader opened a wrapped sleeve of 20 seals and showed the engineer an uneven loading

Most stuck seals were stuck in an alignment as shown:

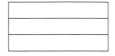

However, some were received stuck in an off-set condition as shown below:

This caused sticking and prevented positioning causing jam-ups.

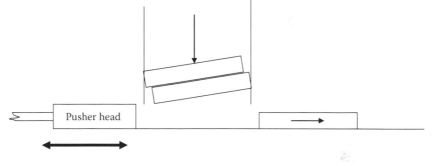

Figure 10.8 Seal alignment and transfer.

condition, as the seals appeared to be stuck together before being unpacked. Most seals were stuck together in multiples of two, three, four, and five (Figure 10.8).

When asked what was done with the jammed-up parts, the loader indicated that he put them on the side and fed them in at a later date as the loader emptied. Based on this information a critical review of the loader operation was conducted, as indicated in Figures 10.9 and 10.10.

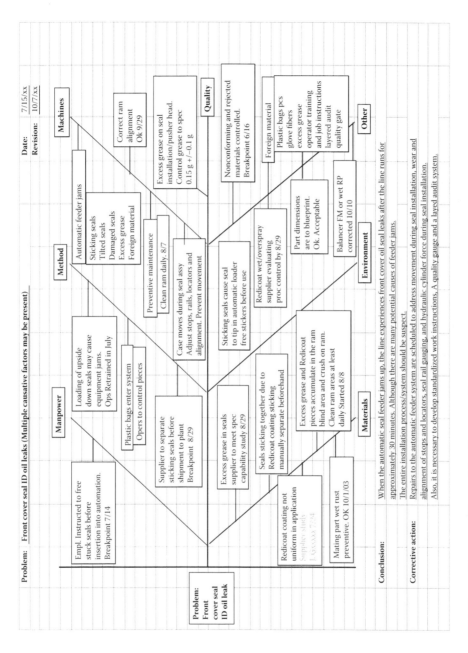

Figure 10.9 Cause diagram of the seal-loading operation.

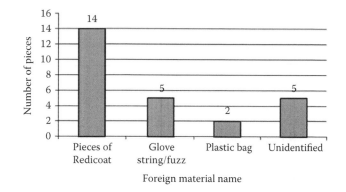

Figure 10.10 Chart of foreign material found in leaking seals (sample of 26, August 6).

This study led to many insights that had not been recognized before. The seal automation loading system was a new installation and had not had the proper evaluation before being put into use. Each of the adverse conditions recognized at that time were consolidated into a corrective action audit sheet, the foremost being the tipped seals that caused the jams (Figure 10.11).

The study of the loading station revealed that there were many other potential different materials that could have caused a jam at the indexer station. The majority of the samples contained shaved pieces of the seal OD readicoat. There were what appeared to be glove strings and fuzzy

- Seals involved in jams are to be sequestered. _____
- There are seals on the floor at the loader area. _____
- Seals are not in an offset condition. _____
- Seals are not stuck together when received. _____
- Front-seal loading area to be cleaned daily. _____
- Loader oil bottle is to be full and operating. _____
- Plastic gloves to be used to load the system. _____
- Auto loaders are not to be loaded above the top. _____
- Plastic wrapper pieces are not present in seal silos. _____
- The seal grease pattern is undamaged and constant. _____
- The seal index pusher is operating smoothly. _____
- There is no other foreign material observable. _____
- Standard work instructions are posted on site. _____
- The loader on site has received instructive training. _____
- Seal coating is cured and dried to prevent hand-ups. _____

Auditor: _____ **Date:** _____

Figure 10.11 Seal audit conditions established.

material that were suspected as originating at both the supplier's and the customer's plants. There were also pieces of plastic bag that may have originated at the in-house seal-loading station. Finally, there was some unidentified material that could not be assigned a cause. The supplier was notified of all conditions and was required to formulate their solutions to the coating and foreign materials problem and to conduct similar audits, as in Figure 10.11.

Front cover oil seal leak: Audit

The following conditions were established and audited.

1. Standardized work instructions are to be followed.
2. Loader employees and relief employees are to be trained.
3. Audit excessive front cover seal leaks (>5 per shift).
4. Establish a layered audit that will check for the following conditions on a routine basis or whenever the quality check exceeds the bogey (>5 per shift).

Obviously, the supplier had a problem with the seals they were delivering. Something in their process had changed that caused them to deliver seals that had an incorrectly cured OD coating. This may have also been the cause of the offset condition of the packaged seals that was causing a jam at the loader indexing site.

This leads us to return to the initial point of this chapter. The three steps to ensure acceptable outputs were not carried at the time. All of the study graphs, illustrations, charts, production piece losses, and dollar losses due to this condition were experienced unnecessarily. If the loader had been instructed and trained to conform to checking the inputs, maintaining quality, and holding flaws, this entire problem would have been avoided. Coincidentally, an audit established in the planning process may have recognized some of the conditions to be controlled.

Checking the inputs would have caused the loader to notify the supervisor, who could have notified and worked with the supplier to correct the seal coating being provided. *Maintaining quality* would have prevented the use of the defective seals. *Holding all flaws* would have prevented the damaged seals that had jammed at the indexer being reloaded into the indexer and being sent to the automatic seal installation station.

O-ring problem

Another quick example of the need to use these simple methods was experienced on an engine assembly line. This operation involved installing a supplier-provided oil dipstick into engines that were being assembled.

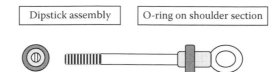

Figure 10.12 Oil dip stick assembly.

The dipsticks were accompanied with an O-ring that provided an air seal at the point of installation into the receiving tube. Upon installation and removal from the engine, it was noticed that the O-ring split and fell off of the dipstick assembly (Figure 10.12).

This sketch depicts the oil dipstick assembly as received from the supplier. The dipstick was composed of a length of metal that had been encapsulated on one end with a plastic handle and shoulder. The shoulder section had a groove in which an O-ring could be retained.

Upon reviewing the defective assemblies it was found that there was a condition that caused the O-ring to tear when the dipstick was inserted and removed from the measuring point. The poor condition was created by a fin on the shoulder section that was caused by a pattern mismatch. The mismatch resulted in the formation of a sharp fin on the shoulder sections, which cut into the O-ring when inserted into the measurement tube. This resulted in torn O-rings and created a product do hold of the engines in the plant, in the warehouse, and in shipment to the assembly plant. The plastic mold is shown in Figure 10.13.

Now, this appears to be a similar problem to one previously discussed. The supplier molding the plastic did not have an effective protective system for their customer. The operator that formed the plastic molded section around the metal dipstick indicator made a defective product. The operators would have been aware of the problem if they had looked at the work that they were performing. This operation violated the *do not create* instruction. In addition, the supplier operator and the final inspection personnel did not react to the defective operation and therefore *did not hold* the defective product; they forwarded the part to the customer. The customer is not removed from blame, as their dipstick installer did not do a visual check on the parts that they were installing into the assembled

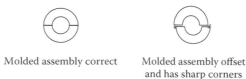

Molded assembly correct Molded assembly offset and has sharp corners

Figure 10.13 Mismatch of two mating plastic sections showing fins.

engine. This violated the instruction to *check* the parts that you are about to use.

I believe that any of these three positions could have easily prevented the problem from occurring if even only one of them complied with these easily applied restrictions.

Summary

This chapter reviewed some manufacturing problems that involved excessive effort to arrive at a solution. It revealed that a lack of simple control procedures by internal or external suppliers to the production line can cause manufacturing problems.

The importance of this type of control is defined by the examples given. They indicate the simplicity of training suppliers to comply with simple control procedures rather than suffer the losses that they do not prevent. Preclusion is much more preferable to a reactive action.

If prevention is required of suppliers, as with the three simple steps of compliance, it will allow inexperienced or inept suppliers to improve their service and their problem-solving abilities. It demonstrates why defective materials should not be accepted at a job site; should not be created by machining, forming, or assembly; and should not be passed on to the next operation or customer. In fact, even competent, experienced suppliers could benefit from establishing these easily installed practices.

The next chapter will discuss the potential to allow the customer to discover weaknesses and to mandate improvements in supplier systems before and after finalization of the supply contract. It is quite lengthy as it provides a focused approach to evaluating the potential effectiveness of a new supplier or the prowess of a supplier undergoing problems.

chapter eleven

System evaluation

The previous chapter studied some problematic manufacturing conditions that required insight into the circumstances that allowed their occurrence. If adverse conditions are recognized beforehand it is possible to prevent their incidence without undue difficulty. The recognition of negative circumstances and the application of corrective actions can help inside or outside suppliers to prevent problems.

Prevention, then, is a preferred action of control to improve the supplier base. Prevention steps can be accomplished by using the three steps of compliance previously discussed but worth mentioning again.

1. Do Check.
2. Don't Create.
3. Do Hold (i.e., sequester).

Unfortunately, there are numerous occasions when people involved with quality improvement within the customer or supplier organization lack these and other tools that are necessary to solve problems. Since both the internal and supplier employees are also providers to the next operation, it is beneficial to have these tools available for their designated use when required. It may even be necessary to mentor suppliers who are having difficulty to improve the supply system and the operating bottom-line financials. Any improvement that can be ascribed to the suppliers, within or outside the customer's organization, can have an impact on the operations involved.

This chapter also relates other important aspects of prevention that can be used to improve the supplier base. Again, this is true whether the supplier is internal or external. These tools provide many deliverables that aid prevention if they are applied consistently.

Deliverables

Deliverables are those benefits that can be attributed to recognized and accomplished changes to the operating system. The benefits can be viewed as improvements that require attainment in order to increase efficiencies or to realize transcendental operations. For example, the deliverables that can be achieved through the use of audits, whether internal or external, are significant. These deliverables can be described as follows:

- System procedures can be documented.
- The timeliness of audit involvement can be defined.
- Audits can produce documented records.
- Audit records indicate compliance or noncompliance.
- Records can be posted and reviewed.
- Key control elements can be listed and defined.
- Deviations from the plan can be observed.
- Corrective actions can be specified and endorsed.
- Corrective actions can be verified.
- New problem conditions can be added to the audit.
- Historical records and improvements can be displayed.
- Audits can be reviewed for revisions to conditions.
- A basis for an applicable improvement file is provided.
- A basis for corrective action plans is provided.
- The presence of a currently operating system is indicated.
- Management are provided with a review mechanism.
- The need for multilayered audits is established.

The list can go on and on, as it is dependent on the operation or service being audited. There are many beneficial results that can be achieved by employing an audit system to improve the supplier base.

As you can observe from the deliverables list, much pertinent information can be captured for future reference and use. However, the deliverables are not the only benefits that can be obtained. Each audit, irrespective of the area of concern, provides a check of the job requirements specific to that area or operation. Each of the observations included as a requirement demonstrates the scrutiny necessary to ensure correct activities.

It really doesn't matter whether an audit of an area is designed for adherence to a specific safety procedure or specific manual steps or motions to complete a repair or a function. An example would be the requirement to wear safety glasses at all times in a designated area, or conversely, the design requirements for conducting an acceptable statistical analysis. Each audit contains pertinent information that is mandated to prevent a recognized potential condition or problem.

For that reason, a specific audit item must include those recognized circumstances that can affect a defined condition. Consequently, it is not difficult to design a preliminary audit to initiate supplier improvements in areas that require adjustment. But each audit must be tailored to meet the conditions encountered for specific locations, jobs, operations, events, or recognized hazards.

Now we must consider the requirements that allow the audit process to be completed.

1. First and foremost, any problem that requires auditing and control should be observed. This might be the recognition of an internal problem or the result of a customer complaint.
2. There should be a listing of a condition or action that has been spontaneously observed by the auditor. This condition or action should be a suspected cause of the problem or anomaly. There should be a means to designate an observable condition as acceptable, unacceptable, or uncertain.
3. There should be a designated area to install condensed notes to ensure that questionable items are fully recognized and reacted to in the near term.
4. The audit form should include instructions and any training that is deemed appropriate to ensure the competency of the auditor.
5. The form must include areas to record the date and name of the person conducting the audit. It should also include the frequency of the audit being conducted.
6. The form should also be used to record any perceived or recognized corrective actions that should be taken. This information could be written or sketched on the reverse side of the form.
7. Audits should be conducted as living documents. Any time a condition is recognized as being inconsequential it should be removed from the listing. Conversely, whenever a condition is recognized that can influence or cause an additional problem it should be added to the listing. This consideration is also applicable to the frequency of audits. It may be advantageous to audit more frequently in times of duress than in times of compliance.

I have provided a plethora of examples that can be used as a basis for initiating this improvement at supplier or customer locations. As pertinent conditions are recognized within your facility or at your supplier, they can be added to the audit list to facilitate supplier improvement.

It matters little if the person establishing an audit has not had expert training in this type of action. Rather, it is the experience gained through auditing that develops the ability to observe actionable items. These items are then recognized and dealt with to prevent an adverse condition from occurring again. For example:

- Are people wearing their safety equipment?
- Are employees using the correct tools for the job?
- Are all gauges in calibration?
- Are the correct functions being performed?

- Are parts removed from the system properly tagged?
- Are parts being perused before use?
- Is the work performed of an acceptable quality?
- Are defective parts being captured?
- Are bad parts being placed in a lockbox?
- Are unnecessary tools being stored on the job site?
- Are job instructions posted and available at the workstation?
- Why did a bad part transfer to this workstation?
- How did a bad part transfer to this workstation?
- Are all the parts on this site in the correct path or routing?

These and many other simple questions can be used to establish audits to correct unfavorable conditions. Initial audit examples are provided in the figures that follow. These example sheets may not contain the exact wording that you may require for auditing your specific areas. However, depending on the terms used within your immediate and or supplier facility, they are proposed as the initial audits of a corrective action program. Additions and deletions can be made at your discretion to each of the following charts (Figures 11.1 through 11.28).

Audit charts

Essentially, the tool to create an audit for any facility is provided by design and process planning and the inputs generated when problems are encountered. By observing the problems that are reported and by proposing viable circumstances, an audit can be created as a means of process control. Most times, an audit will prove beneficial to the prevention of problems from its initiation. At other times, the number of specific causative factors may be too numerous to generate at one observation or specification setting.

Remember the causes of mixed parts that were discussed in a previous chapter. The initial audits were revised numerous times until what was deemed to be the current prevention list was developed, notwithstanding the numerous other potential causes for mixed parts that can be added to the considerations for an audit list.

But an effective audit must start somewhere. The sooner a problem is recognized and named, the sooner it can be defined and eliminated. Failure to recognize and name a problem only increases the time required to respond and to find a suitable corrective action. These audits have proven to be exceedingly beneficial in the improvement of internal and external suppliers. These improvements have resulted in considerable cost savings and process improvements for both the supplier and the customer.

Continuous work	Weekly			Monthly			Variable			Date: Auditor 1: Auditor 2:
Response: Y = Compliant, N = Noncompliant, and ? indicates an observation in question. Place an "X" in box Y if compliant or an "X" in box N if noncompliant. Place an "X" in box ? if unsure.	Y	N	?	Y	N	?	Y	N	?	List corrective actions on reverse side! Briefly list discrepancy or condition of concern
1	Work procedure documented?									
2	Work instructions available to Applicable workforce?									
3	Work instructions posted?									
4	Do operators have formal training?									
5	Is training certified and recorded?									
6	All can read/understand postings?									
7	Are instructions proceduralized?									
8	Revisions require approvals?									
9										
10										
11										
12										
13										
14										
15										
16										

Figure 11.1 Continuous work area audit.

Die casting	Weekly			Monthly			Variable			Date: Auditor 1: _____ Auditor 2: _____
Response: Y = Compliant, N = Noncompliant, and ? indicates an observation in question.	Y	N	?	Y	N	?	Y	N	?	List corrective actions on reverse side!
Place an "X" in box Y if compliant or an "X" in box N if noncompliant. Place an "X" in box ? if unsure.										Briefly list discrepancy or condition of concern
1	Is area clean?									
2	Work instructions posted?									
3	Start-up approvals given?									
4	Alluminum alloy in spec?									
5	Shop order correct?									
6	PM sheet up to date?									
7	Setup approval is OK?									
8	Hot oil at temperature?									
9	Monitor system checked?									
10	Records are complete?									
11	Foreign material controlled?									
12	Checksheets used and maintained?									
13	Checksheets show recent activity?									
14	Procedures followed?									
15	Safety equipment used?									
16	Any other concerns?									

Figure 11.2 Die casting department audit.

Die and pattern inspection	Weekly			Monthly			Variable			Date: Auditor 1: _____ Auditor 2: _____
Response: Y = Compliant, N = Noncompliant, and ? indicates an observation in question.	Y	N	?	Y	N	?	Y	N	?	**List corrective actions on reverse side!**
Place an "X" in box Y if compliant or an "X" in box N if noncompliant. Place an "X" in box ? if unsure.										**Briefly list discrepancy or condition of concern**
1 Is area clean?										
2 Dies, patterns, and serials identified?										
3 Storage is arranged?										
4 Job instructions posted?										
5 Patterns stored properly?										
6 Gauge verification sheets being maintained?										
7 All gauges identified and calibrations current?										
8 Is there a gauge calibration procedure in effect?										
9 Gauge check-in/check-out Procedure being used?										
10 Checksheets being used?										
11 Are checksheets being completed?										
12 All procedures being followed?										
13 Any other concerns?										
14										

Figure 11.3 Die and pattern inspection area audit.

Final Inspection	Weekly			Monthly			Variable			Date: _____ Auditor 1: _____
Response: Y = Compliant, N = Noncompliant, and ? indicates an observation in question.	Y	N	?	Y	N	?	Y	N	?	List corrective actions on reverse side!
Place an "X" in box Y if compliant or an "X" in box N if noncompliant. Place an "X" in box ? if unsure.										Briefly list discrepancy or condition of concern
1 Inspection area is clean?										
2 All gauges have current calibration?										
3 Work areas are neat and orderly?										
4 If SPC charts are being used, do they have correct control limits?										
5 Are blueprints up to date with correct revisions being used?										
6 Are job instructions posted?										
7 Do the inspectors understand the job instructions?										
8 Are current customer problems displayed and identified?										
9 Final sampling plan being used?										
10 Is the Sampling Plan approved and applicable for the right confidence?										
11 Are checksheets being used?										
12 Are checksheets complete?										
13 Defect control procedure present?										
14 Are all procedures being followed?										
15 Any other visible concerns?										

Figure 11.4 Final inspection area audit.

Foreign material control	Weekly			Monthly			Variable			Date: _____ Auditor 1: _____ Auditor 2: _____
Response: Y= Compliant, N = Noncompliant, and ? indicates an observation in question.	Y	N	?	Y	N	?	Y	N	?	**List corrective actions on reverse side!**
Place an "X" in box Y if compliant or an "X" in box N if noncompliant. Place an "X" in box ? if unsure.										**Briefly list discrepancy or condition of concern**
1	Housekeeping audits conducted?									
2	Procedures are in effect?									
3	Containers inspected before use?									
4	Containers returned from suppliers to be perused for foreign material and cleanliness?									
5	Equipment purged and cleaned between batches?									
6	Routine maintenance inspections conducted on equipment?									
7	Proper disposal of materials evaluated on regular basis?									
8	FM receptacles made available throughout the facility?									
9	FM receptacles are emptied?									
10	Old tags removed from storage tubs and containers?									
11	No other concerns?									
12										
13										
14										
15										
16										

Figure 11.5 Foreign material control audit.

Gauge storage		Weekly			Monthly			Variable			Date: Auditor 1: Auditor 2:
		Y	N	?	Y	N	?	Y	N	?	List corrective actions on reverse side!
Gauge room												Briefly list discrepancy or condition of concern
Place an "X" in box Y if compliant or an "X" in box N if noncompliant. Place an "X" in box ? if unsure.												
1	Area is clean?											
2	Temperature controlled?											
3	Humidity controlled?											
4	Calibration results available?											
5	Are calibration records timely and complete?											
6	Job instructions posted?											
7	All inspectors certified to perform calibrations?											
8	All gauges clearly identified?											
9	Is there a gauge calibration procedure in effect?											
10	Gauge check-in/check-out procedure being used>											
11	Checksheets being used?											
12	Checksheets complete?											
13	Defect control procedure present?											
14	All procedures being followed?											
15	Any other visible concerns?											

Figure 11.6 Gauge storage area audit.

Hazardous waste	Weekly			Monthly			Variable			Date: Auditor 1: Auditor 2:
Response: Y = Compliant, N = Noncompliant, and ? indicates an observation in question.	Y	N	?	Y	N	?	Y	N	?	**List corrective actions on reverse side!**
Place an "X" in box Y if compliant or an "X" in box N if noncompliant. Place an "X" in box ? if unsure.										**Briefly list discrepancy or condition of concern**
1 Is area clean?										
2 Is area dry?										
3 All containers labeled?										
4 Is area gated and locked?										
5 No sewer grates nearby?										
6 Daily inspections made?										
7 No leaking containers?										
8 Timely permits posted?										
9 Attending employees trained?										
10 Records are complete?										
11 In compliance with governmental and universal standards?										
12 Checksheets used and maintained?										
13 Checksheets show recent activity?										
14 Procedures followed?										
15 Safety equipment used?										
16 Any other concerns?										

Figure 11.7 Hazardous waste area audit.

Individual workstations	Weekly			Monthly			Variable			Date: Auditor 1: ___ Auditor 2: ___
	Y	N	?	Y	N	?	Y	N	?	**List corrective actions on reverse side!**
Response: Y = Compliant, N = Noncompliant, and ? indicates an observation in question.										
Place an "X" in box Y if compliant or an "X" in box N if noncompliant. Place an "X" in box ? if unsure.										**Briefly list discrepancy or condition of concern**
1 Area is clean?										
2 Area is well lighted?										
3 Work areas are neat and orderly?										
4 Gauges match job order?										
5 Unnecessary gauges not present?										
6 Job instructions are posted?										
7 Job instructions are understood?										
8 Understands *do check, don't make,* and *do hold* rules?										
9 Foreign material controlled?										
10 Checksheets are used and maintained?										
11 Checksheets show activity?										
12 Procedures followed?										
13 Safety equipment used?										
14 Any other concerns?										

Figure 11.8 Individual workstation audit.

Layered audits	Weekly			Monthly			Variable			Date: _____ Auditor 1: _____ Auditor 2: _____
Response: Y = Compliant, N = Noncompliant, and ? indicates an observation in question. Place an "X" in box Y if compliant or an "X" in box N if noncompliant. Place an "X" in box ? if unsure.	Y	N	?	Y	N	?	Y	N	?	**List corrective actions on reverse side!**
										Briefly list discrepancy or condition of concern
1 Audit procedure exists?										
2 Audit frequency followed?										
3 All people levels involved?										
4 Are deviations noted and corrective actions taken?										
5 Preventive corrections and Actions are employed?										
6 Are adjustments made to DFMEA, PFMEA, control plan, and work instructions?										
7 Audits reviewed by upper management?										
8 Any other concerns?										
9										
10										
11										
12										
13										
14										
15										
16										

Figure 11.9 Layered audit.

Lessons learned	Weekly			Monthly			Variable			Date: Auditor 1: _____ Auditor 2: _____
Response: Y = Compliant, N = Noncompliant, and? indicates an observation in question.	Y	N	?	Y	N	?	Y	N	?	**List corrective actions on reverse side!**
Place an "X" in box Y if compliant or an "X" in box N if noncompliant. Place an "X" in box ? if unsure.										**Briefly list discrepancy or condition of concern**
1 There is a procedure to capture information?										
2 Procedure is documented and shows current use?										
3 Information is relayed to sister functions?										
4 Improvements are provided to concerned individuals?										
5 Adjustments are made to DFMEA, PFMEA, control plan, and work instructions?										
6 Audits reviewed by upper management?										
7 No other concerns?										
8										
9										
10										
11										

Figure 11.10 Lessons learned audit.

Machining area	Weekly			Monthly			Variable			Date: _____ Auditor 1: _____ Auditor 2: _____
	Y	N	?	Y	N	?	Y	N	?	**List corrective actions on reverse side!**
Response: Y = Compliant, N = Noncompliant, and ? indicates an observation in question.										
Place an "X" in box Y if compliant or an "X" in box N if noncompliant. Place an "X" in box ? if unsure.										**Briefly list discrepancy or condition of concern**
1 Machining area is clean?										
2 All gauges have current calibration?										
3 Work areas are neat and orderly?										
4 Gauges match job order?										
5 Unnecessary gauges not present?										
6 Coordinate measuring machine programs at latest revision?										
7 Operators understand job detail?										
8 Work instructions posted?										
9 Are checksheets being used?										
10 Are checksheets complete?										
11 If SPC tools are being used, do they have correct control limits?										
12 Is there a control procedure for machining defects?										
13 Are lockboxes used for control?										
14 Are all procedures being followed?										
15 Safety equipment being used?										
16 Are there other visible concerns?										

Figure 11.11 Machining area audit.

Maintenance	Weekly Y N ?	Monthly Y N ?	Variable Y N ?	List corrective actions on reverse side! Briefly list discrepancy or condition of concern
Response: Y = Compliant, N = Noncompliant, and ? indicates an observation in question. Place an "X" in box Y if compliant or an "X" in box N if noncompliant. Place an "X" in box ? if unsure.				
1 Maintenance area is clean?				
2 Preventive maintenance in use?				
3 Plant lubrication scheduled?				
4 Work areas neat and orderly?				
5 Emergency systems tested?				
6 Emergency testing records current and posted?				
7 Alarms are on, functional and egularly audited?				
8 Fire control sprinkling systems under pressure?				
9 Plant air pressure operates within control limits?				
10 Cooling tower water temperature is controlled to a limit?				
11 Are checksheets being used?				
12 Are checksheets complete?				
13 Safety lockout procedure audited on a regular basis?				
14 Automatic safety lockout Devices tested on a regular basis?				
15 Safety equipment being used?				
16 Are there other visible concerns?				

Date: _____
Auditor 1: _____
Auditor 2: _____

Figure 11.12 Maintenance area audit.

Metal melting		Weekly			Monthly			Variable			Date: _____ Auditor 1: _____ Auditor 2: _____
Response: Y = Compliant, N = Noncompliant, and ? indicates an observation in question.		Y	N	?	Y	N	?	Y	N	?	**List corrective actions on reverse side!**
Place an "X" in box Y if compliant or an "X" in box N if noncompliant. Place an "X" in box ? if unsure.											**Briefly list discrepancy or condition of concern**
1	Is area clean?										
2	Ladles maintained?										
3	Furnaces maintained?										
4	Transfer equipment maintained?										
5	Chemistry acceptable?										
6	Temperature audited?										
7	Temperatures controlled?										
8	Scales calibrated?										
9	Ingots color coded?										
10	All metals identified?										
11	Foreign material controlled?										
12	Checksheets being used?										
13	Checksheets maintained?										
14	Procedures being followed?										
15	Safety equipment used?										
16	Any other concerns?										

Figure 11.13 Metal-melting area audit.

Mistake prevention	Weekly			Monthly			Variable			Date: ____ Auditor 1: ____ Auditor 2: ____
Response: Y = Compliant, N = Noncompliant, and ? indicates an observation in question.	Y	N	?	Y	N	?	Y	N	?	**List corrective actions on reverse side!**
Place an "X" in box Y if compliant or an "X" in box N if noncompliant. Place an "X" in box ? if unsure.										**Briefly list discrepancy or condition of concern**
1	Are there lockboxes used for containing defectives?									
2	Are all prevention devices listed?									
3	Are all prevention devices tested?									
4	Procedure is present for testing the prevention devices?									
5	Are all devices designed to be fail safe?									
6	Can devices be bypassed?									
7	Plans to continue production are available for device failure?									
8	Employees trained in use of the protective devices?									
9	Employee training documented?									
10	No other concerns?									

Figure 11.14 Mistake prevention audit.

Nonconforming materials	Weekly			Monthly			Variable			Date: _____ Auditor 1: _____ Auditor 2: _____
	Y	N	?	Y	N	?	Y	N	?	**List corrective actions on reverse side!**
Response: Y = Compliant, N = Noncompliant, and ? indicates an observation in question. Place an "X" in box Y if compliant or an "X" in box N if noncompliant. Place an "X" in box ? if unsure.										**Briefly list discrepancy or condition of concern**
1 A procedure is available?										
2 Procedure is being followed?										
3 Suspect materials immediately tagged and sequestered?										
4 People peruse materials before use?										
5 People held to correct operations?										
6 People hold questionable parts?										
7 Photos or samples describe acceptable components as may be applicable?										
8 Posted instructions direct employee action when in question?										
9 All can understand instructions?										
10 Only customer-approved repairs are applied to reworked product?										
11 Parts use response time criteria?										
12 No other concerns?										
13										
14										

Figure 11.15 Nonconforming materials audit.

Part traceability	Now			Daily			Variable			Date: Auditor 1: _____ Auditor 2: _____
										List corrective actions on reverse side!
Response: Y = Compliant, N = Noncompliant, and ? indicates an observation in question.	Y	N	?	Y	N	?	Y	N	?	
Place an "X" in box Y if compliant or an "X" in box N if noncompliant. Place an "X" in box ? if unsure.										**Briefly list discrepancy or condition of concern**
1　Is there a tracking procedure?										
2　Are date codes applicable?										
3　Are pattern serials identified?										
4　Can parts be traced back to their originating components or raw materials?										
5　No outside laboratory analysis will be required for raw materials?										
6　Materials will not require certifications from other source?										
7　Contracts specify no change in material without authorization?										
8　No other concerns?										
9										
10										
11										
12										

Figure 11.16　Part traceability audit.

Problem-solving acumen		Weekly			Monthly			Variable			Date: Auditor 1: Auditor 2:
Response: Y = Compliant, N = Noncompliant, and ? indicates an observation in question. Place an "X" in box Y if compliant or an "X" in box N if noncompliant. Place an "X" in box ? if unsure.		Y	N	?	Y	N	?	Y	N	?	**List corrective actions on reverse side!** **Briefly list discrepancy or condition of concern**
1	Someone assigned to solve specified problems?										
2	Team approach used to define problem statement?										
3	Problems assigned by financial impact?										
4	Problems fully defined?										
5	Measuring system discrete?										
6	Visual observations made?										
7	Experiments conducted?										
8	Root causes verified?										
9	Corrective actions tested?										
10	Corrective actions recorded?										
11	DFMEA adjusted?										
12	PFMEA adjusted?										
13	Control plan adjusted?										
14	Work instructions adjusted?										
15	Posted instructions adjusted?										
16	Workforce trained?										

Figure 11.17 Problem-solving acumen audit.

Process capability	Now			Daily			Variable			Date: _____ Auditor 1: _____ Auditor 2: _____ **List corrective actions on reverse side!** **Briefly list discrepancy or condition of concern**
Response: Y = Compliant, N = Noncompliant, and ? indicates an observation in question. Place an "X" in box Y if compliant or an "X" in box N if noncompliant. Place an "X" in box ? if unsure.	Y	N	?	Y	N	?	Y	N	?	
1 Capability an issue? (If not an issue then go to next sheet.)										
2 Is there a capability procedure?										
3 All critical dimensions and characteristics have an acceptable CPK >1.33?										
4 Supplier understands the critical dimensions that require acceptable statistical control?										
5 CPK calculations are verified with acceptable current analysis?										
6 Is there a plan to improve noncompliant capability?										
7 Is sorting to be used to meet specifications until capability can be achieved?										
8 Is there assurance that sorting will be effective?										
9 Will sorting be done by manual or mechanized means?										
10										

Figure 11.18 Process capability audit.

Date: _____
Auditor 1: _____
Auditor 2: _____

Process improvements	Weekly			Monthly			Variable			List corrective actions on reverse side! Briefly list discrepancy or condition of concern
Response: Y = Compliant, N = Noncompliant, and ? indicates an observation in question. Place an "X" in box Y if compliant or an "X" in box N if noncompliant. Place an "X" in box ? if unsure.	Y	N	?	Y	N	?	Y	N	?	
1 Multiple function team evaluates and establishes PFMEA?										
2 PFMEA includes safety issues?										
3 PFMEA includes rework units?										
4 PFMEA includes exception work?										
5 PFMEA improvements are addressed and ongoing?										
6 PFMEA improvements are assigned by impact severity?										
7 PFMEA items are assigned to individuals for correction?										
8 PFMEA items contain a due date for completion?										
9 PFMEA llists temporary actions to overcome temporary difficulties?										
10 All PFMEA revisions are approved by the customer before installation?										
11 There is evidence of ongoing PFMEA improvements?										
12 No other concerns?										

Figure 11.19 Process improvement audit.

Production approval	Weekly			Monthly			Variable			Date: Auditor 1: Auditor 2:
Response: Y = Compliant, N = Noncompliant, and ? indicates an observation in question.	Y	N	?	Y	N	?	Y	N	?	**List corrective actions on reverse side!**
Place an "X" in box Y if compliant or an "X" in box N if noncompliant. Place an "X" in box ? if unsure.										**Briefly list discrepancy or condition of concern**
1	Is there a meeting for design function and production?									
2	Is a design checklist used and completed for product?									
3	Are preproduction process plans developed and approved?									
4	Are DFMEA processes used?									
5	Are PFMEA processes used?									
6	Evidence of corrective actions is shown on current PFMEAs?									
7	Do these functions meet the required supplier standards?									
8	Are lockboxes considered for use to control defective material?									
9	Any other visible concerns?									
10										

Figure 11.20 Production approval audit.

Quality assurance areas	Weekly			Monthly			Variable			Date: _____ Auditor 1: _____ Auditor 2: _____
Response: Y = Compliant, N = Noncompliant and ? indicates an observation in question. Place an "X" in box Y if compliant or an "X" in box N if noncompliant. Place an "X" in box ? if unsure.	Y	N	?	Y	N	?	Y	N	?	LIST CORRECTIVE ACTIONS ON REVERSE SIDE! Briefly list discrepancy or condition of concern
1 Inspection area is clean?										
2 All gauges have current calibration?										
3 Work areas are neat and orderly?										
4 If SPC charts are being used, do they have correct control limits?										
5 Are blueprints up to date with correct revisions being used?										
6 Are coordinate measuring machine programs up to the latest revision?										
7 Are current customer problems displayed and identified?										
8 Are first piece inspection parts tagged and identified?										
9 Are job instructions posted?										
10 Do the quality assurance staff understand the job instructions?										
11 Are checksheets being used?										
12 Are checksheets complete?										
13 A control procedure for defects?										
14 Are all procedures being followed?										
15 Safety equipment being used?										
16 Are there other visible concerns?										

Figure 11.21 Quality assurance area audit.

Date: _____
Auditor 1: _____
Auditor 2: _____

Response time	Now			Daily			Variable			List corrective actions on reverse side!
Response: Y = Compliant, N = Noncompliant, and ? indicates an observation in question. Place an "X" in box Y if compliant or an "X" in box N if noncompliant. Place an "X" in box ? if unsure.	Y	N	?	Y	N	?	Y	N	?	Briefly list discrepancy or condition of concern
1 Response procedure stated?										
2 Individual assignmments made?										
3 Team activity is mandated with responsible individuals?										
4 List of appropriate actions reviewed with each problem?										
5 At a minimum product is held at: plant, dock, shipping, trucking, warehouse, etc.?										
6 Problem product contained as per established list?										
7 All customer disruptions and sequesters are scrutinized?										
8 Customer(s) are immediately notified of sequestration?										
9 Customer(s) are updated on status of problem daily?										
10 DFMEA, PFMEA, control plans, work instructions, and applicable postings are corrected?										
11 No other concerns?										

Figure 11.22 Response time audit.

Shipping areas	Weekly			Monthly			Variable			Date: Auditor 1: Auditor 2:
Response: Y = Compliant, N = Noncompliant, and ? indicates an observation in question.	Y	N	?	Y	N	?	Y	N	?	**List corrective actions on reverse side!**
Place an "X" in box Y if compliant or an "X" in box N if noncompliant. Place an "X" in box ? if unsure.										**Briefly list discrepancy or condition of concern**
1	Is area clean?									
2	Shipping labels attached properly?									
3	Parts have required part number?									
4	Labels filled in completely?									
5	Containers have approved authorizations?									
6	Parts banded as required?									
7	Parts shrink-wrapped as required?									
8	Scales calibrated?									
9	Foreign material present?									
10	Checksheets being used?									
11	Checksheets maintained?									
12	Procedures being followed?									
13	Safety equipment in use?									
14	Any other concerns?									

Figure 11.23 Shipping area audit.

Tool room		Weekly			Monthly			Variable			Date: _____ Auditor 1: _____ Auditor 2: _____
											List corrective actions on reverse side!
Response: Y = Compliant, N = Noncompliant, and ? indicates an observation in question.		Y	N	?	Y	N	?	Y	N	?	Briefly list discrepancy or condition of concern
Place an "X" in box Y if compliant or an "X" in box N if noncompliant. Place an "X" in box ? if unsure.											
1	Area is clean?										
2	Temperature controlled?										
3	Humidity controlled?										
4	Calibration results available?										
5	Are calibration records timely and complete?										
6	Job instructions posted?										
7	All personnel certified to perform calibrations?										
8	All gauges clearly identified?										
9	Is there a mold or pattern approval process in effect and being used?										
10	Gauge check-in/check-out procedure being used?										
11	Checksheets being used?										
12	Checksheets complete?										
13	Defect control procedure present?										
14	All procedures being followed?										
15	Any other visible concerns?										

Figure 11.24 Tool room audit.

Tier 2 supplier support	Weekly			Monthly			Variable			Date: Auditor 1: Auditor 2:
Response: Y = Compliant, N = Noncompliant, and ? indicates an observation in question.	Y	N	?	Y	N	?	Y	N	?	**List corrective actions on reverse side!**
Place an "X" in box Y if compliant or an "X" in box N if noncompliant. Place an "X" in box ? if unsure.										**Briefly list discrepancy or condition of concern**
1	Supplier directs their suppliers?									
2	This supplier requires Tier 2 suppliers to conform to their requirements?									
3	Supplier aids Tier 2 suppliers in problem solving?									
4	Supplier aids Tier 2 suppliers in developing a continuous improvement system?									
5	Supplier demands that Tier 2 suppliers conform to *no change without preapproval?*									
6	No other concerns?									
7										
8										

Figure 11.25 Tier 2 supplier support audit.

Unauthorized changes	Weekly			Monthly			Variable			Date: ___ Auditor 1: ___ Auditor 2: ___
Response: Y = Compliant, N = Noncompliant, and ? indicates an observation in question.	Y	N	?	Y	N	?	Y	N	?	List corrective actions on reverse side!
Place an "X" in box Y if compliant or an "X" in box N if noncompliant. Place an "X" in box ? if unsure.										Briefly list discrepancy or condition of concern
1 Change procedure exists?										
2 No changes made without customer preapproval?										
3 Customer to approve any plans for proposed changes?										
4 Authorized changes require bank-sized protection?										
5 Failed authorized trial changes have backup protection?										
6 Supplier's supplier(s) aware of unauthorized change rules?										
7 Supplier's supplier(s) aware that inconsequential change approval is required?										
8 No other concerns?										
9										
10										

Figure 11.26 Unauthorized changes audit.

Workstation	Daily		Weekly/Tier		Variable		Date: _____ Auditor 1: _____ Auditor 2: _____
Response: Y = Compliant, N = Noncompliant, and ? indicates an observation in question.	Y	N ?	Y	N ?	Y	N ?	**List corrective actions on reverse side!**
Place an "X" in box Y if compliant or an "X" in box N if noncompliant. Place an "X" in box ? if unsure.							**Briefly list discrepancy or condition of concern**
1	There is adequate lighting?						
2	Workbenches are adjustable?						
3	Fixtures used for assembly?						
4	Jigs used for assembly?						
5	Designated tools stored in designated areas?						
6	Only designated gauges stored in designated spaces?						
7	Job instructions posted?						
8	Photos or sketches of work to be accomplished posted?						
9	No ergonomic problems?						
10	Rubber mat on concrete floor?						
11	Operators can stop process?						
12	Lights or bells used to communicate troubles?						
13	Environment-friendly?						
14	No other concerns?						

Figure 11.27 Workstation audit.

General	Weekly			Monthly			Variable			Date: Auditor 1: Auditor 2:
Response: Y = Compliant, N = Noncompliant, and ? indicates an observation in question.	Y	N	?	Y	N	?	Y	N	?	**List corrective actions on reverse side!**
Place an "X" in box Y if compliant or an "X" in box N if noncompliant. Place an "X" in box ? if unsure.										**Briefly list discrepancy or condition of concern**
1 There is a preventive maintenance program for equipment?										
2 All stored and processing product is identified?										
3 Areas do not contain spilled loads or scattered material?										
4 A lubrication policy is in effect and is current?										
5 Clean room cleanliness standards are in effect?										
6 Finished product is not allowed to sit in a negative environment?										
7 Current quality levels and problems are posted?										
8 Job instructions are posted at repetitive operation stations?										
9 Employees follow *do check, don't make and do hold policy?*										
10										

Figure 11.28 General area audit.

Summary

This chapter proposed a significant amount of information that can be used and applied to improve supplier bases. The information is applicable to upstream operations, downstream operations, and external operations.

Audits can be instrumental in preventing antagonistic conditions that can arise if harmful conditions are not recognized and controlled. Audits can elicit deliverables that will benefit the user or the mentor that applies them. The use of these area-specific tools can result in process improvements, scrap reduction, rework or repair reduction, increased profits, greater customer satisfaction, and safer work practices to name a few considerations. It befits all supervisors and managers to recognize the importance of applying audits to correct troublesome conditions internally and externally.

The next chapter follows the supplier enhancement sequence. Some of the previous chapters have discussed some of the supplier problems that are prevalent in industry today. Other chapters have listed information that will aid suppliers to perform at a higher level of competence. The following chapter will discuss those considerations that are necessary for developing and applying a supplier's corrective action plan. These considerations might also be applied to the evaluation of proposed suppliers who have not have been previously utilized.

chapter twelve

Other requirements

The previous chapter provided an accumulation of sample audits that can be applied to different areas, circumstances, or work operations. They were introduced to familiarize the reader with the ease with which they can be developed to mentor inside and outside suppliers. If audits are performed on supplier problem conditions they will result in improvements in the supply system as corrective actions are applied.

Now, having recognized that a problem exists with a current supplier, it is necessary to implement actions that will prevent that same problem from occurring again. This chapter attempts to define those specific criteria that should be demanded of suppliers when manufacturing or service problems arise. It also contains references to some conditions that must be considered prior to contract approval for new suppliers that are about to be included in the tiered system.

Corrective action criteria

There are many areas of importance in a supplier's corrective action plan. Initially, corrective action plans for problems internal to the customer are specifically controlled by the application of internal procedures. These procedures depend on the expertise of the organization itself.

Outside suppliers are subject to mentoring to achieve acceptable performance levels, as deemed necessary by the customer. However, the customer has the task of judging the importance of each criteria that they may demand. In some cases, the following areas of constraint are recognized as achievable and controllable to ensure quality supplies. The improvement procedure may require some if not all of these requirements.

Required compliance

Compliance to customer requirements is deemed necessary in at least four specific areas.

1. Documentation of the quality system
2. Implementation of the quality plan
3. Availability of records and information
4. Management participation

These areas, if under control and active participation, can provide the basis for problem elimination and process improvement.

Documentation of the quality system

The specifics for documentation are related to the quality system that is prevalent and being utilized by individual customers. It may differ slightly from industry to industry, but there are common concerns that must be addressed. The required documentation contains information that gives the reviewer an indication of the competence of the supplying organization.

Generally, the documentation features an acceptable process flow diagram covering all operations, including the following areas:

- Receiving dock activities
- Receiving activities
- Operation descriptions
- Refurbishment operations
- Label information
- Inspection operations
- Shipping activities
- Capturing scrap
- Hazardous waste control
- Lessons learned

A second form of documentation is a process control plan that contains comparable information, as does the PFMEA. It includes relevant values and describes any critical product characteristics that must be controlled by gauging and specified sample sizes at designated frequencies. The document must suggest the actions necessary to address those items of high concern in the PFMEA.

The third form of required documentation is a completed PFMEA that is deemed acceptable for the work described. It must contain relevant information that adequately describes all of the measurable components and must be treated as a living document that shows current revisions or corrections.

These three documents provide evidence that the process has been subject to a thorough analysis that is composed of a product flow, process controls, and a failure mode analysis specific to the current system. These documents also allow for continual improvement activities to be incorporated and balanced to prevent problem recurrence.

Implementation of the quality plan

After it is determined that there is a documented quality system, it is necessary to determine if the plan can be effectively applied. To begin with, each industry may have different requirements, but there are common threads that can be considered to determine the plan's adequacy.

Some considerations are as follows:

- Do job stations match the process flow plan?
- Is the work place arrangement reasonable?
- Are job instructions available to allow referral?
- Do operators understand the instructions?
- Are operators permitted to curtail problematic operations?
- Does the process control plan match the control plan?
- Are the gauges calibrated and identified with instructions?
- Are the gauges, jigs, and fixtures subject to a control plan?
- Do operators visually peruse incoming materials?
- Do operators know what is critical to control?
- Can operators stop production when parts are defective?
- Do operators hold parts deemed questionable?
- Are questionable parts tagged and placed in a lockbox?
- Do operators perform the work as per the instructions?
- Do inspectors know what is most critical to control?
- Are inspections performed as per the requirements?
- Is the process capable as described and practiced?
- If statistical process control (SPC) is used, are charts viable and show control?
- Are completed parts protected from damage?

These observations can be conducted by means of a plant visit and a preliminary audit of conditions at the supplier. But again, these attributes should also be checked at the customer's location to ensure that they are also attempting to protect their customer(s).

Availability of records and information

The process of keeping accurate records is a requirement that can help to identify problems should they occur. The records are not qualified by their length. Rather, their importance is determined by the information that they contain. Some considerations in the record and information areas are as follows:

- Has the supplier been involved in the DFMEA process?
- Does the supplier have knowledge of the DFMEA process?
- Has the supplier had access to the DFMEA to develop the PFMEA?
- Does the supplier have the approved drawings required?
- Do all the supplier departments involved have the same drawings available?
- Are the drawings at the correct change level?

- Does the supplier understand the critical dimensions and requirements that are mandatory?
- Are the drawings complete?
- Does the supplier agree to the *No change without prior approval* policy that will be in effect?

Once these records and agreements are in place there should be no reason for changes to be made without prior approval from the customer. Suppliers that do not or cannot adhere to this requirement should be replaced by a competent rival.

Management participation

The importance of the inclusion of management in the quality plan cannot be underemphasized. The prominence of maintaining adequate control of the quality system is essential to the success of both the supplier and the customer.

Management actions

To ascertain whether management is involved in the quality system it is essential that specific actions be observed. These actions include the study and recognition of problem conditions that can be related to inadequacies present in the system. Questions that can address the adequacy of management involvement are as follows:

- Are problem conditions recognized?
- Are corrective actions made immediately?
- Can operators stop a defective process?
- If not the operator, who can intervene?
- Are out-of-control conditions corrected?
- Do action plans specify corrective actions?
- Are control and improvement plans followed?
- Is there acceptable communication between operators?
- Is there acceptable communication between shifts?
- Is there a qualified designated individual involved?
- Are scrap, rework, and downtime charts displayed?
- Are examples of current problems on display?
- Is there a message board for current problems?
- Are customer problems recorded and analyzed?
- Are refurbishment and reinspection costs scrutinized?
- Are emergency and premium freight costs examined?
- Have the PFMEA, control plans, and instructions been revised to address recent problems?

As you may have noticed, management involvement is also essential to assure compliance with the customer requirements. These management participation activities in conjunction with the creation of documentation, activity implementation, and the availability of records and information can provide the basis for a *lessons learned* file. The lessons learned can be applied to the customer internally or to numerous other tiered supplier organizations.

New contract requirements

There are some conditions that must be considered prior to contract approval for suppliers that are about to be included in the tier system. Since most suppliers exhibit some euphoria when a lucrative contract is finalized, some have a tendency to overstate their capability to the potential customer. Because of this overenthusiasm it is advisory to perform a system of checks and balances to ensure that the supplier can provide what they promise.

It is best to set preconditions for the supplier and mentor them in the evaluative process so that they may in turn mentor their suppliers. This passing on of knowledge and requirements will provide a basis for preventing unwarranted surprises once a production schedule is established. Since not all suppliers are aware of the conditions that may be necessary to provide an acceptable service, this task of studying the supplier base should be conducted by qualified personnel from the originating customer. This is especially true if the suppliers that will be involved lack various QS or ISO certifications.

The requirements of contract completion should include some basic considerations at a minimum. The considerations for workstation operations must be spelled out. These could include the use of the tools, methods, inspections, and tests to control the quality output of the supplier base.

Workstation operations

Supplier workstation considerations that are deemed necessary include the following concerns:

- A process flow diagram is developed and available.
- Boundary samples should be available for review.
- Job instructions should be available and posted.
- Operators should have undergone training.
- Operators should be aware of the potential problems.
- Rework operations should be defined.
- Layouts should include the manufacturing flow.
- Sorting activities should be defined.
- Do hold or lockbox areas should be defined.
- A tour of the proposed facilities should be conducted.

Inclusion of these job site attributes will improve the quality and performance of work performed at the individual workstations.

Quality process pointers

The following points should be considered to determine the supplier's compliance with the quality process.

- Ensure that historical data indicates an acceptable quality level as defined by parts per million (PPM) or defects per million opportunities (DPMO).
- Confirm that provisions are in place to track the PPM or DPMO performance of the proposed parts after start-up.
- Determine the scope and use of the error-proofing devices and activities that are proposed.
- Determine if activities are subject to error proofing with the use of masks, barriers, tests, or other devices.
- Determine the actions to be taken to correct the top five PFMEA high scores to ascertain the supplier's intentions to reduce the characteristic impact by reducing the risk priority number (RPN)-scored value on the PFMEA.
- Review the control plan to judge its appropriateness.
- Decide if the functional tests proposed will result in adequate results.

These features directly affect the quality and quantity of the output that will be provided.

Manufacturing metrics

The requirements to consider when evaluating the viability of proposed supplier manufacturing metrics are as follows:

- Does the facility have the capacity to accept new work?
- Has the supplier calculated and projected the number of completed jobs per hour?
- Does the supplier operate on one or two shifts?
- Can new work be scheduled into available shift time?
- Is weekend scheduling applicable and appropriate?
- Are there capacity concerns that exceed availability?
- Are new equipment or facility increases required?
- Can new equipment be obtained in time to achieve the schedule?
- Is the projected installation time adequate to allow schedule compliance?
- Is adequate time provided for an equipment debug operation?

- Does the supplier possess the capacity for potential schedule increases?
- Does the supplier possess proof of achieving their prior manufacturing metric estimates?

Answering these questions will provide insight into potential problems that will affect the manufacturing schedule.

Compliance and performance readiness

The following questions will determine the readiness of the supplier to actively supply the product under consideration.

- Does the supplier meet all the requirements of the preproduction approval process? That is, does the supplier have all the basic qualifications that are required by the customer to meet a mandatory preproduction approval process?
- Is supplier compliant with the most current design status?
- Does the supplier possess the production approval released by the customer?
- What is the status of the supplier contract?
- Has the supplier shown competence in achieving a minimal production run?
- Does the supplier have the capacity to produce the product in the amount required?
- Do the supplier's past actions indicate that they have control of their suppliers to prevent unauthorized changes?

These considerations give an indication of the current readiness of the supplier to begin supplying the sample or preproduction parts for approval. If they don't comply, then remedial actions are required.

Operational issues

The list of items that can affect operations include the following precautions, which can impact a customer organization if it is not aware of the consequences.

- Are there any manpower concerns?
- Are there trained individuals available to sustain production and any required schedule increases?
- Will all supplier product be supplied from the same prequalified location and process?
- Can part traceability be provided from start to finish?

- Do throughput concerns exist that affect the product handling or flow?
- Is a preventive maintenance (PM) program available?
- Does the PM program show up to date actions?
- Is the PM program functional?
- Is the supplier's supplier up to speed with the requirements being imposed by the customer?
- What Tier 2 supplier does the Tier 1 supplier intend to use for the entire contract period?
- Will all major issues be rectified before a scheduled preproduction meeting?
- Has the supplier exhibited the ability to respond to changing conditions to meet increasing schedules?
- Has the supplier provided any major concerns to be addressed?

These issues will instill a degree of confidence that any operational issues that can be focused on or prevented will not affect the supplier's ability to meet the production schedule.

Raw and finished material metrics

The following minimum requirements are necessary to meet the manufacturing desires of some large customer organizations.

- Is the raw material source capable and viable?
- Can schedules be verified via visual representations?
- Can the supplier receive shipment instructions via electronic means rather than by fax?
- Are flats, trays, and designed containers available as per the production schedule?
- Is the supplier aware of the identification and traceability requirements?
- Is the supplier aware of the need to provide laboratory certifications for all materials to ensure process capability, as may be required for silicones, etc.
- Does the supplier know where product can be marked for identification so as not to appear on the finished product after assembly?
- Can shipments be impacted by imposed border inspections, delays, or hang-ups?
- Can the supplier relate to and provide a plan to enable shipments to overcome schedule inhibitors?
- Does the supplier have the capacity to meet the schedule?
- Has the supplier addressed dunnage concerns?
- Has the supplier addressed transportation concerns?

- Has the supplier addressed supply chain concerns?
- What other concerns are noteworthy?

Addressing these concerns will help to ensure that the raw and finished material metrics are fulfilled.

Plant visits and tours

In addition to all of these requirements, the customer may insist on a plant visit to observe the conditions under which the final work will be performed. This is especially critical when reviewing a supplier's capability if no previous experience has been established.

Some of the concerns raised during this visit are included in the previous list, and others are developed by visual inspection of the facility. They may include

- Protections established to prevent mixed parts
- Provisions established to control foreign materials
- Posting job instructions at the workstations
- Observing operators working in compliance with posted work instructions
- The availability of boundary samples for review at the workstations
- The presence of lockboxes to contain materials considered defective
- Observing operators to determine if the rules *Do Check*, *Don't Create*, and *Do Hold* are being employed to control defective materials
- The control of product lots or batches
- A containment plan

The plant visit is essential and can be accomplished in a relatively short period of time if the supplier is within a moderate distance. More distant suppliers can be visited by contracted outside sources that are available within their location boundaries.

Summary

This chapter addressed many concerns that should be considered when applying an internal or external corrective action plan to a supplier. The actions proposed are a guide to prevent current problems from occurring again after corrective actions have been established.

Knowing the specific criteria to demand from suppliers is important when evaluating the future compliance of the supplying concern. These considerations result in the prevention of problems if the concerns are addressed before the contract is finalized. Many problems are the result of seemingly insignificant oversights that allow malfunctions to persist.

To achieve acceptability the supplier needs to comply with the customer's requirements.

1. The quality plan must be documented.
2. The quality plan must be implemented.
3. Records and information must be available.
4. Management must participate.

These considerations should also be applied to the evaluation of proposed suppliers who have not been used previously. And it should also be applied to any Tier 2 suppliers that may be involved in the supply chain process. In addition to the established criteria, special attention must be directed to evaluating new suppliers in the following areas of concern:

1. New contract requirements
2. Workstation operations
3. Quality process pointers
4. Manufacturing metrics
5. Compliance and performance readiness
6. Operational issues
7. Raw and finished material metrics
8. Plant visits and tours

It is important that experienced suppliers mentor their Tier 2 supplier base to conform to the requirements of the customer. This activity will result in untold benefits for all the organizations involved.

Prevention steps have proven effective in reducing costs, improving quality, and reducing the manufacturing and service problems experienced in the supply chain.

The next chapter contains information relevant to problem solving that may not be part of the supplier base's knowledge. It will facilitate problem solving by presenting specific examples that can be applied by any supplier that may be experiencing manufacturing difficulties. Customers have found this information to be effective in improving results when mentoring and aiding their suppliers to improve their problem-solving activities. It is proposed as a means of prevention to deter production upsets due to the supplier's lack of problem-solving experience.

chapter thirteen

Tools for improvement

The previous chapter dealt with requirements that make the facilitation of corrective changes easier. These practices include establishing criteria for corrective action, compliance to customer requirements, the documentation and use of the quality system, the availability of records and information, management participation and action, new contract requirements, workstation observation, quality process pointers, manufacturing metrics, readiness, operational issues, and finally, plant visits and tours of the facilities.

This chapter deals with a more pragmatic topic: the issue that not all suppliers are equipped to solve supplier and manufacturing problems efficiently. The examples and illustrations will provide a few meaningful solutions to supplier manufacturing and quality problems. These methods are to be used in conjunction with the customer-supplied mentoring process. In addition, there is a method proposed to ensure that a problem is corrected once the internal or external supplier has indicated that corrective changes have been applied.

The improvement process is depicted in Figure 13.1.

The chapter then deals with a few useful tools that may not be available to some suppliers. In fact, the people involved with quality improvement within a customer or supplier organization often lack the necessary skills to solve these problems. Since both these internal and external supplier employees are also providers to the next operation, it is beneficial for these tools to be available for their designated use.

This chapter provides a couple of easily applied problem-solving methods that can enhance a supplier's performance. The first deals with answering a few simple questions to identify potential problem areas. After identifying causative factors, solutions can be considered and applied. The second method provided can be used as an aid to help an internal or external supplier to properly identify which part of a process is more significantly affecting their output when schedules cannot be maintained. These conditions that cause downtime or rework are compared with those conditions that create scrap or other detrimental inactivity. These methods have been applied effectively to various industries.

Both of these considerations will be addressed. The informing and mentoring of suppliers need not be complicated. In some cases it may require the presentation of only a few examples that can be used as

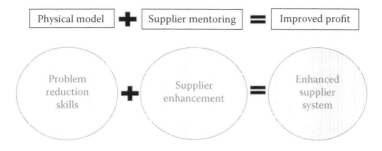

Figure 13.1 Improved profitability with supplier mentoring.

references when a similar problem is experienced. Hopefully, these examples will provide the suppliers with a basic technical ability to analyze and identify the cause of the problem. Once recognized, the problem can be defined and the corrective actions applied to enable improvements in the supplier's process.

It befits the customer to aid the supplier in their problem elimination endeavors for their mutual benefit. This mentoring can be accomplished via actual meetings or conferencing with suppliers at the customer's or supplier's location when dire problems arise. It is not unusual when working with suppliers to require specific actions on their part while attempting to analyze and correct production problems. To augment that requirement, it seems clear that the customer should help the supplier if they do not have the analytic experience necessary to provide important information.

Answering the right questions

In previous chapters we viewed a couple of individual analyses of problems that were being evaluated. They required much thought and time but did not result in any meaningful recommendations or results due to supplier inexperience.

In contrast, observe the following analysis, which is much more definitive (Figure 13.2). It provides specific actionable items for resolution. The analysis attempted to determine why electronic chips from the supplier failed a final inspection and test after assembly.

However, even this improvement in analysis is lacking in specificity as to what really caused the problem within the chip manufacturer's facility. It really only addresses why the problem occurred, which is only one-third of an appropriate analysis. In addition to determining why the problem occurred it is necessary to evaluate two other criteria: the *Why Sequence*.

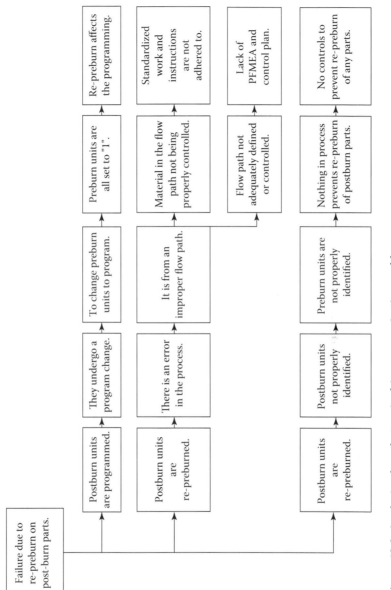

Figure 13.2 Analysis of an electronic chip communication problem.

The other two criteria to be addressed are

1. Why the problem was not detected
2. Why the problem was not predicted

These two considerations will not be addressed individually as they can each be analyzed in the same manner as was used to address why the original problem occurred. Remember when analyzing a problem it is wise to address all three criteria so that similar problems can be prevented by making corrective changes to the manufacturing, the quality, and the planning disciplines that may also be involved.

I believe that these two simple tools will aid problem prevention and may help to enhance your supplier's problem-solving skills. Therefore, it is essential that problem prevention be considered in the DFMEA, the PFMEA, and the control plan and in error-proofing activities when products or services are being developed.

The next tool is not as simple but it is just as critical. Its use may be required by an internal or external supplier when they have difficulty providing the necessary parts required. This method can enhance a supplier's ability to achieve and maintain their production schedule. To provide a better understanding of the problem it is necessary to provide some background to the circumstances involved.

A means for achieving production schedules

A customer had contracted a supplier to provide three sizes of machined flywheels to fulfill their production needs. The supplier had one machine that could perform the operation. The machining operation for each of the three flywheels required that it pass through three operations on the same machine. There were no other machines or processes available where the operations could be performed. When changing from one of the flywheels to another the machine had to undergo a major changeover to facilitate the machining requirements.

The customer did not adequately assess the supplier's ability to handle the production schedule before the contract was finalized. After providing a sufficient number of parts for start-up, the schedule was increased as production was ramped up. It soon became clear that the supplier was not able to maintain the required production schedule.

An immediate problem-solving activity was created to resolve the problem. The supplier was required to provide an updated copy of their PFMEA to determine the most significant risk priority number (RPN) at present. (An RPN is the result of a mathematical comparison that indicates the effect of risk on production and quality.) It was believed that an analysis of the supplier's PFMEA might provide an indication of areas where potential improvements could be made.

The supplier complied with the request and provided the document shown in partial form in Figures 13.3 and 13.4. (The right-hand side of the document is excluded from the illustration as it does not add any relevant information or explanation.)

Process function	Potential failure mode	Potential failure effect	S E V	Potential cause/ mechanism for failure	O C C	Current process control	D E T	R P N	Action
Stage 1	O/S U/S OD	Can't chuck at next station	4	Worn insert	3	Tool change schedule	3	36	
			1	Maladjustment	3	Tool change schedule	3	9	
	O/S U/S thickness	Affects final dimensions	2	Broken insert	2	Tool change schedule	3	12	
			2	Misread gauge	2	Tool change schedule	3	12	
		No clean-up later	2	Broken insert	2	Tool change schedule	3	12	
	O/S U/S chamfer	Change volume	1	Broken insert	1	Station 3	3	3	
			1	Maladjustment	5	Station 3	4	20	
Stage 2	O/S U/S Bore	Can't chuck at next station	5	Maladjustment	3	Operator inspection	5	75	
		Hit Metal at balance system	2	Broken insert	2	Next station	5	20	
		No clean-up at balance station	2	Broken insert	2	Operator inspection	2	6	
	O/S U/S thickness	Affects final dimensions	5	Broken insert	3	Operator inspection	10	150	
	O/S U/S chamfer	Binder	1	Maladjustment	2	Operator inspection	3	6	
		Change volume	1	Maladjustment	2	Operator inspection	3	6	
		Change bearing area	1	Maladjustment	2	Operator inspection	3	6	
	Concentricity	Out of round	3	Maladjustment	6	Operator inspection	3	54	
		Out of balance	3	Maladjustment	6	Operator inspection	3	54	
	Bore roundness	Out of balance	3	Material stress	1	First piece inspection	2	6	
		Out of concentricity	3	Machine bearing wear	1	First piece inspection	2	6	

Figure 13.3 Page 1 of the flywheel PFMEA.

Process function	Potential failure mode	Potential failure effect	SEV	Potential cause/mechanism for failure	OCC	Current process control	DET	RPN	Action
Stage 3	O/S U/S OD	Change volume	1	Worn insert	3	First piece inspection	4	12	
		Binder	7	Broken insert at stage 1	3	First piece inspection	3	63	
		Damping	8	Adjustment	5	First piece inspection	3	120	
	O/S U/S bearing thickness	No clean-up at balance system	4	Stage 2 worn insert	3	First piece inspection	3	36	
	O/S U/S bearing step	Hit metal at balance system	3	Stage 2 broken insert	4	First piece inspection	3	36	
		Damping	8	Setup tooling	3	First piece inspection	4	96	
		Change volume	2	Worn/broken insert	4	First piece inspection	1	8	
	Sides perpendicular to bore	Balance affected	8	Machining	6	First piece inspection	5	240	
	Parallelism	Balance affected	8	Nib extension	1	First piece inspection	10	80	
		Chamfer	2	Chamfer at station 2	1	First piece inspection	10	20	
	O/S U/S step diameter	Nylon masking problem	2	Adjustments	3	First piece inspection	1	6	
	Finish	Coating adhesion	1	Feed is too slow	1	First piece inspection	1	1	

Figure 13.4 Page 2 of the flywheel PFMEA.

Unfortunately, the supplier was inexperienced in the formation and application of the PFMEA instrument. However, they did attempt to analyze their operation to the best of their ability. As you can see, there are potential problems with their analysis. There are no indications of corrective action plans to address the high RPNs on the list to help their production process. (RPN = severity rating × occurrence severity × detection severity; or S × O × D.)

Document provided for flywheels

The form does indicate, however, that there are three stations recognized, and their potential RPNs are as follows: Station 1: 104; Station 2: 389; Station 3: 718.

So, the data indicates that Station 3 has the highest RPN, which signifies that it could have the greatest impact on the manufacturing and quality process. (The later in a process stream a problem is found, the more costly it is to discover it after all the resources have been invested.) Therefore, any problem found at Station 3 would more significantly affect the production output than those found at Station 1 or 2. Any of the conditions that were reflected by a large RPN number would reduce the potential output of the machining operation.

Causes of failure

From this data it was possible to determine the effects of failure to ascertain what could be done to improve the output. To accomplish this it was necessary to determine the causes of failure provided by the RPN numbers generated. The causes of failure were as follows (Figure 13.5).

As can be observed from Figure 13.5, there is an excessive negative effect on the supplier's manufacturing process due to worn and broken inserts that require a substantial amount of adjustments to be made. (RPN = 362 + 266 + 128 = 756; this represented 756/1217 or 62.1% of the potential loss of production attributed to worn or broken inserts.) The supplier immediately contacted the supplier of their machining tools and was able to secure inserts that were longer living and more resistant to breakage.

New cutting tools proved to be beneficial, and the supplier was able to increase their production moderately. But they were still unable to meet the projected schedule increases that were forecast. This is the major point that I want to illuminate for your future consideration.

The supplier and the customer were interested in increasing production to meet the requirements. They attempted to do this by focusing on items that interfered with the current process. They did not focus on another aspect that was more critical to this problem: those practices that physically stopped the process. These practices only consisted of adjusting for the broken inserts and the machine setup time to change from one flywheel model to another. Broken inserts resulted in an RPN value of 266, whereas the machine setup required a complete production stop for up to eight hours for each change. The machine setup time successfully stopped the process for almost the entire production shift. The frequency of the

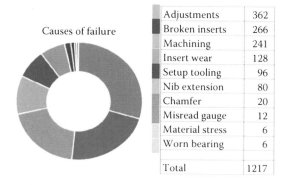

Causes of failure		
	Adjustments	362
	Broken inserts	266
	Machining	241
	Insert wear	128
	Setup tooling	96
	Nib extension	80
	Chamfer	20
	Misread gauge	12
	Material stress	6
	Worn bearing	6
	Total	1217

Figure 13.5 RPN values for flywheel failures.

practice was established in an attempt to control inventories, prevent scrap, and improve material turnover rates to reflect lean practices.

Flywheel final resolution

The ability to change equipment frequently and to transfer minimal lot sizes is a recognized means of lean manufacturing. But in this case a prerequisite of the equipment changes was not adhered to. The supplier should have recognized the lack of production caused by the delay due to inefficient machine setup methods. The supplier lacked the foresight to assign a team to address the need to improve the time allotted for equipment changes. Right from the start, unsupervised setups were allowing excessive time for the changes. As a matter of fact, the original change and setup time was reduced to four hours when it was recognized and placed under closer supervision. This 50% improvement resulted from the immediate gain per equipment change and setup, which occurred three times per week. Without any additional improvements, there was a gain of 12 hours of usable production per week. This halting of production was at least equal to all of the other more significant problems that only degraded the process (Figure 13.6).

This was not the culmination of the study. The supplier was encouraged to assign an improvement team to study and reduce the time required to change and setup the equipment, as they were required to increase production once again to meet the projected schedule. The supplier realized that there was potential for improvement when mentored to improve their

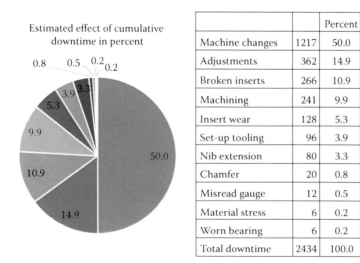

Estimated effect of cumulative downtime in percent			Percent
Machine changes	1217	50.0
Adjustments	362	14.9
Broken inserts	266	10.9
Machining	241	9.9
Insert wear	128	5.3
Set-up tooling	96	3.9
Nib extension	80	3.3
Chamfer	20	0.8
Misread gauge	12	0.5
Material stress	6	0.2
Worn bearing	6	0.2
Total downtime	2434	100.0

Figure 13.6 Effect of equipment changes and problems.

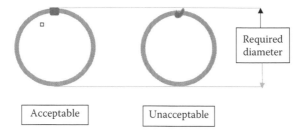

Figure 13.7 Sketch of fuel pipe condition.

operations. The elimination of *dead time* when no production can be achieved is to be addressed whenever a scheduling problem arises. It can aid the supplier, whether internal or external, to achieve the requirements imposed by the customer. Therefore, dead time should be considered to be any condition that causes the machine to be idle that can be eliminated or minimized.

Finally, here is a proposed method to improve the customer's confidence in the supplier's corrective actions. It is best illustrated with an example. Suppose that a fuel line supplier to the automotive industry has shipped leaking fuel lines to a customer. Let's say it was determined that the leak was caused by a poor weld and an undersized outside diameter that subtracted from the circularity (roundness) of the pipe that fit into a connection (Figure 13.7).

All that needs to be accomplished is for the customer to request some data from the supplier. Have the supplier make a comparison between five newly produced parts chosen at random versus five parts that were made before the corrective changes were made. Measure them to the criteria and plot them in a table, as shown in Figure 13.8.

Dimension	Pair 1		Pair 2		Pair 3		Pair 4		Pair 5	
	New	Old	New	Old	New	Old	New	Old	New	Old
Smooth acceptable weld within 0.002 in.	Yes	Yes	Yes	No	Yes	No	Yes	Yes	Yes	Yes
Acceptable diameter within 0.002 in.	Yes	No	Yes	No	Yes	No	Yes	No	Yes	No

Figure 13.8 Table of comparison.

The results of the comparison of the randomly chosen new parts with the old production parts reveals the following:

1. The change does give a degree of confidence that the weld changes that were made are acceptable because all five of the newly produced parts meet the 0.002-inch tolerance specified, compared with only three of the five old welds. However, there may be a tie or overlap in the third pair of data, which precludes acceptance without comparing the old unit in pair 3 to the new units in pairs 1, 2, 4, and 5 for clear data separation.
2. The revised diameter size indicates that the changes made were acceptable when the five new randomly chosen parts were measured to the tolerance specified, whereas the older parts did not meet the tolerance. There is clear separation here, so it can be accepted with 95% confidence that the change was effective and that the new parts meet the specifications.

Now, this is not to say that there is certainty in the results of the changes, but there is in excess of 95% confidence that the result is accurate. This is because there are only 252 ways of combining five good (new) and five bad (old) samples. In addition, there are only 8 possible results out of the 252 combinations where the complete separation of good from bad results in a condition of seven or more and where there is no overlap in values (conditions). That is, at least seven of the values are not tied or mixed. So, if all the new diameter values are correct and meet specifications, whereas all the old ones do not, then there is a significant change present at 95% confidence.

Summary

A couple of problem-solving methods were provided that can enhance a supplier's performance. This need not be a complicated process that entails extreme effort. Rather, the supposition is that a supplier can improve their capabilities by recognizing that there should be at least three questions asked when a problem occurs:

1. Why did the problem occur?
2. Why was the problem not detected?
3. Why was the problem not predicted?

It takes these three questions to recognize the current adversity. Suppliers should ask what can be done to highlight a problem if it appears and what changes have to be made to the planning process so that all three of these questions are considered in the manufacture of a future product. So, answering a few basic questions is one of the things to take away from this section.

The second thing to take away is to consider the impact of conditions that completely stop the supplier manufacturing process (dead time). In most cases, the application of a study team to reduce the time to change equipment or dies is not considered a means of meeting a production schedule. Rather, some suppliers will consider running a process through lunch or other breaks. With other suppliers the tendency may be to add additional shifts, overtime, or weekend work to augment the rate of production.

If equipment is subject to long changeover periods, efforts can be made to minimize these practices, as explained in the sample. It should be considered whether improvements can be made without requiring additional manpower or overtime to meet a production schedule.

Finally, a comparison of previous production with current production using random samples can indicate with a degree of confidence the date where at least seven values are not comingled.

The next chapter will present some easily applied audit forms that can be used at internal or external supplier locations to improve their operation performance.

chapter fourteen

Useful audit criteria

Chapter 13 provided a couple of recommendations to aid the supplier in achieving their production schedule efficiently. These involve predicting and safeguarding against problems so that corrective actions can prevent their happening again. This can require some deep thought to understand conditions that may not be known. It also provided a little insight into how to improve productive output and how to determine if changes are effective.

This section requires minimal education, understanding, or skill. It contains samples of audits that can be used by anyone interested in understanding how easily problems can be recognized with little effort. It can also be used to ensure that prescribed actions are being utilized. The chapter contains 28 worksheets that can be used as audits by the reader to assess a supplier's activity. They are applicable to both internal and external suppliers.

I recommend that you consider the following proposal. Having read the book, consider the audits to be tools to improve the internal or external supplier base. Use the audits as a starting point to assess supplier compliance. Start within your own organization and pick a function within the audit sheets provided. Or make up your own initial form and use it on a specific area of your concern. Refer to the worksheets that follow (Figures 14.1 through 14.28).

Example audit considerations

Summary

Merely take a walk through the department under question and look to see what is happening. But, before taking the walk, review one of the department sheets to understand what you are looking to investigate. It shouldn't take more than a minute or two for you to appreciate the conditions under review and to evaluate current practices.

As you complete your walk, notice the conditions present. If you are unsure of a compliance, ask someone a specific question. If you find something that troubles you or an operation that isn't on the audit, make a mental note. Finally, stop to record your observations. If you have found everything to be satisfactory and in compliance, that's great, but hardly likely! If you find items not in compliance or unsatisfactory or

Continuous work		Date: _____ Auditor: _____
Audit daily		List corrective actions on reverse side
Items to be audited as applicable		Briefly list discrepancy or condition of concern
1	Work procedure documented?	
2	Work instructions available to applicable workforce?	
3	Work instructions posted?	
4	Do operators have formal training?	
5	Is training certified and recorded?	
6	All can read/understand postings?	
7	Are instructions proceduralized?	
8	Revisions require approvals?	
9		
10		
11		
12		
13		
14		
15		
16		

Figure 14.1 Continuous work audit sheet.

questionable, you have identified an auditable item. If you recognize a required condition that is not on the audit sheet, add it to the sheet for the next audit. Fill out the audit form and assign a champion to correct all adverse items observed. Assign someone to conduct an audit at irregular times to ensure that all corrections and gains can be maintained.

These, then, are usable audits worksheets that can be used as a base for the improvement of suppliers whether they be internal or external to the parent organization. They have been proven to be useful and easily applicable to the specified activities, controls, and methods required to provide a satisfactory product or service. The can also be used to recognize other potentially harmful activities that can be added to a revised audit form. They should be included in a continuous improvement program to aid both internal and external suppliers.

Die casting	Date: _____ Auditor: _____	
Audit daily	**List corrective actions on reverse side**
Items to be audited as applicaable	**Briefly list discrepancy or condition of concern**

	Items to be audited as applicaable	Briefly list discrepancy or condition of concern
1	Is area clean?	
2	Work instructions posted?	
3	Start-up approvals given?	
4	Alluminum alloy in spec?	
5	Shop order correct?	
6	PM sheet up to date?	
7	Setup approval is OK?	
8	Hot oil at temperature?	
9	Monitor system checked?	
10	Records are complete?	
11	Foreign material controlled?	
12	Checksheets used and maintained?	
13	Checksheets show recent activity?	
14	Procedures followed?	
15	Safety equipment used?	
16	Any other concerns?	

Figure 14.2 Die casting audit sheet.

Now, it is obvious that all audits could not be contained in this book because of the many different manufacturers and service suppliers that have sundry operations in various fields. It is now up to the reader to create applicable audits that can be of the most benefit to your organization, enhance suppliers, and improve profits.

The next chapter contains a summary of the information presented in this discourse.

Die and pattern inspection	Date: _____ Auditor: _____
Audit daily	**List corrective actions on reverse side!**
Items to be audited as applicable	**Briefly list discrepancy or condition of concern**
1 Is area clean?	
2 Dies, patterns, and serials identified?	
3 Storage is arranged?	
4 Job instructions posted?	
5 Patterns stored properly?	
6 Gauge verification sheets being maintained?	
7 All gauges identified and calibrations current?	
8 Is there a gauge calibration procedure in effect?	
9 Gauge check-in/check-out procedure being used?	
10 Checksheets being used?	
11 Are checksheets being completed?	
12 All procedures being followed?	
13 Any other concerns?	
14	
15	

Figure 14.3 Die and pattern inspection audit sheet.

Final inspection	Date:
	Auditor:
Audit daily	**List corrective actions on reverse side!**
Items to be audited as applicable.	**Briefly list discrepancy or condition of concern**
1	Inspection area is clean?	
2	All gauges have current calibration?	
3	Work areas are neat and orderty?	
4	If SPC charts are being used, do they have correct control limits?	
5	Are blueprints up to date with correct revisions being used?	
6	Are job instructions posted?	
7	Do the inspectors understand job instructions?	
8	Are current customer problems displayed and identified?	
9	Final sampling plan being used?	
10	Is the sampling plan approved and applicable for the right confidence?	
11	Are checksheets being used?	
12	Are checksheets complete?	
13	Defect control procedure present?	
14	Are all procedures being followed?	
15	Any other visible concerns?	
16		

Figure 14.4 Final inspection audit sheet.

Foreign material control		Date: _____ Auditor: _____
Audit daily		**List corrective actions on reverse side!**
Items to be audited as applicable		**Briefly list discrepancy or condition of concern**
1	Housekeeping audits conducted?	
2	Procedures are in effect?	
3	Containers inspected before use?	
4	Containers returned from suppliers to be perused for foreign material and cleanliness?	
5	Equipment purged and cleaned between batches?	
6	Routine maintenance inspections conducted on equipment?	
7	Proper disposal of materials evaluated on regular basis?	
8	FM receptacles made available throughout the facility?	
9	FM receptacles are emptied?	
10	Old tags removed from storage tubs and containers?	
11	No other concerns?	
12		
13		
14		
15		
16		

Figure 14.5 Foreign material control audit sheet.

Gauge storage	Date: _____ Auditor: _____	
Audit daily	**List corrective actions on reverse side!**	
Items to be audited as applicable.	**Briefly list discrepancy or condition of concern**	
1	Area is clean?	
2	Temperature controlled?	
3	Humidity controlled?	
4	Calibration results available?	
5	Are calibration records timely and complete?	
6	Job instructions posted?	
7	All inspectors certified to perform calibrations?	
8	All gauges clearly identified?	
9	Is there a gauge calibration procedure in effect?	
10	Gauge check-in/check-out procedure being used?	
11	Checksheets being used?	
12	Checksheets complete?	
13	Defect control procedure present?	
14	All procedures being followed?	
15	Any other visible concerns?	

Figure 14.6 Gauge storage audit sheet.

General	Date: _____ Auditor: _____
Audit daily	**List corrective actions on reverse side!**
Items to be audited as applicable.	**Briefly list discrepancy or condition of concern**
1 There is a preventive maintenace program for equipment?	
2 All stored and processing product is identified?	
3 Areas do not contain spilled loads or scattered material?	
4 A lubrication policy is in effect and is current?	
5 Clean room cleanliness standards are in effect?	
6 Finished product is not allowed to sit in a negative environment?	
7 Current quality levels and problems are posted?	
8 Job instructions are posted at repetitive opereation stations?	
9 Employees follow *do check*, *don't make* and *do hold* policy?	
10	
11	
12	
13	
14	
15	

Figure 14.7 General audit sheet.

Hazardous waste	Date: _____ Auditor: _____
Audit daily	**List corrective actions on reverse side!**
Items to be audited as applicable	**Briefly list discrepancy or condition of concern**
1 Is area clean?	
2 Is area dry?	
3 All containers labeled?	
4 Is area gated and locked?	
5 No sewer grates nearby?	
6 Daily inspections made?	
7 No leaking containers?	
8 Timely permits posted?	
9 Attending employees trained?	
10 Records are complete?	
11 In compliance with governmental and universal standards?	
12 Checksheets used and maintained?	
13 Checksheets show recent activity?	
14 Procedures followed?	
15 Safety equipment used?	
16 Any other concerns?	

Figure 14.8 Hazardous waste audit sheet.

	Individual work stations	Date: _____ Auditor: _____
	Audit daily	**List corrective actions on reverse side!**
	Items to be audited as applicable	**Briefly list discrepancy or condition of concern**
1	Area is clean?	
2	Area is well lighted?	
3	Work areas are neat and	
4	Gauges match job order?	
5	Unnecessary gauges not present?	
6	Job instructions are posted?	
7	Job instructions are	
8	Understands *do check*, *don't make*, and *do hold* rules?	
9	Foreign material controlled?	
10	Checksheets used and maintained?	
11	Checksheets show recent activity?	
12	Procedures followed?	
13	Safety equipment used?	
14	Any other concerns?	
15		
16		

Figure 14.9 Individual workstation audit sheet.

Layered audits	Date: _____ Auditor: _____	
Audit daily	**List corrective actions on reverse side!**	
Items to be audited as applicable	**Briefly list discrepancy or condition of concern**	
1	Audit procedure exists?	
2	Audit frequency followed?	
3	All people levels involved?	
4	Are deviations noted and Corrective actions taken?	
5	Preventive corrections and Actions are employed?	
6	Are adjustments made to DFMEA, PFMEA, control plan, and work instructions?	
7	Audits reviewed by upper management?	
8	Any other concerns?	
9		
10		
11		
12		
13		
14		
15		
16		

Figure 14.10 Layered audit worksheet.

Lessons learned		Date: _____
		Auditor: _____
Audit daily		**List corrective actions on reverse side!**
Items to be audited as applicable		**Briefly list discrepancy or condition of concern**
1	There is a procedure to capture information?	
2	Procedure is documented and shows current use?	
3	Information is relayed to sister functions?	
4	Improvements are provided to concerned individuals?	
5	Adjustments are made to DFMEA, PFMEA, control plan, and work instructions?	
6	Audits reviewed by upper management?	
7	No other concerns?	
8		
9		
10		
11		
12		
13		
14		
15		
16		

Figure 14.11 Lessons learned audit sheet.

Machining area	Date: _____ Auditor: _____
Audit daily	**List corrective actions on reverse side!**
Items to be audited as applicable	**Briefly list discrepancy or condition of concern**
1 Machining area is clean?	
2 All gauges have current calibration?	
3 Work areas are neat and orderly?	
4 Gauges match job order?	
5 Unnecessary gauges not present?	
6 Coordinate measuring machine programs at latest revision?	
7 Operators understand job detail?	
8 Work instructions posted?	
9 Are checksheets being used?	
10 Are checksheets complete?	
11 If SPC tools are being used, do they have correct control limits?	
12 Is there a control procedure for machining defects?	
13 Are lockboxes used for control?	
14 Are all procedures being followed?	
15 Safety equipment being used?	
16 Are there other visible concerns?	

Figure 14.12 Machining area audit sheet.

Maintenance		Date: _____ Auditor: _____
Audit daily		**List corrective actions on reverse side!**
Items to be audited as applicable		**Briefly list discrepancy or condition of concern**
1	Maintenance area is clean?	
2	Preventive maintenance in use?	
3	Plant lubrication scheduled?	
4	Work areas neat and orderly?	
5	Emergency systems tested?	
6	Emergency testing records current and posted?	
7	Alarms are on, functional and egularly audited?	
8	Fire control sprinkling systems under pressure?	
9	Plant air pressure operates within control limits?	
10	Cooling tower water temperature is controlled to a limit?	
11	Are checksheets being used?	
12	Are checksheets complete?	
13	Safety lockout procedure audited on a regular basis?	
14	Automatic safety lockout devices tested on a regular basis?	
15	Safety equipment being used?	
16	Are there other visible concerns?	

Figure 14.13 Maintenance audit sheet.

| Metal melting | Date: _____ |
| | Auditor: _____ |

| Audit daily | List corrective actions on reverse side! |
Items to be audited as applicable	Briefly list discrepancy or condition of concern
1 Is area clean?	
2 Ladles maintained?	
3 Furnaces maintained?	
4 Transfer equipment maintained?	
5 Chemistry acceptable?	
6 Temperature audited?	
7 Temperatures controlled?	
8 Scales calibrated?	
9 Ingots color coded?	
10 All metals identified?	
11 Foreign material controlled?	
12 Checksheets being used?	
13 Checksheets maintained?	
14 Procedures being followed?	
15 Safety equipment used?	
16 Any other concerns?	

Figure 14.14 Metal-melting audit sheet.

Mistake prevention	Date: _____ Auditor: _____
Audit daily	**List corrective actions on reverse side!**
Items to be audited as applicable	**Briefly list discrepancy or condition of concern**

	Items to be audited as applicable	Briefly list discrepancy or condition of concern
1	Are there lockboxes used for containing defectives?	
2	Are all prevention devices listed?	
3	Are all prevention devices tested?	
4	Procedure is present for testing the prevention devices?	
5	Are all devices designed to be fail safe?	
6	Can devices be by passed?	
7	Plans to continue production are available for device failure?	
8	Employees trained in use of the protective devices?	
9	Employee training documented?	
10	No other concerns?	
11		
12		
13		
14		
15		
16		

Figure 14.15 Mistake prevention audit sheet.

Nonconforming materials	Date: _____ Auditor: _____
Audit daily	**List corrective actions on reverse side!**
Items to be audited as applicable	**Briefly list discrepancy or condition of concern**

	Items to be audited as applicable	Briefly list discrepancy or condition of concern
1	A procedure is available?	
2	Procedure is being followed?	
3	Suspect materials immediately tagged and sequestered?	
4	People peruse materials before use?	
5	People held to correct operations?	
6	People hold questionable parts?	
7	Photos or samples describe acceptable components as may be applicable?	
8	Posted instructions direct employee action when in question?	
9	All can understand instructions?	
10	Only customer-approved repairs are applied to reworked product?	
11	Parts use response time criteria?	
12	No other concerns?	
13		
14		
15		
16		

Figure 14.16 Nonconforming materials audit sheet.

Part traceability	Date: _____ Auditor: _____
Audit daily	**List corrective actions on reverse side!**
Items to be audited as applicable	**Briefly list discrepancy or condition of concern**
1	Is there a tracking procedure?	
2	Are date codes applicable?	
3	Are pattern serials identified?	
4	Can parts be traced back to their originating components or raw materials?	
5	No outside laboratory analysis will be required for raw materials?	
6	Materials will not require certifications from another source?	
7	Contracts specify no change in material without authorization?	
8	No other concerns?	
9		
10		
11		
12		
13		
14		
15		
16		

Figure 14.17 Part traceability audit sheet.

Problem-solving acumen	Date: _____ Auditor: _____	
Audit daily	**List corrective actions on reverse side!**	
Items to be audited as applicable	**Briefly list discrepancy or condition of concern**	
1	Someone assigned to solve specified problems?	
2	Team approach used to define problem statement?	
3	Problems assigned by financial impact?	
4	Problems fully defined?	
5	Measuring system discrete?	
6	Visual observations made?	
7	Experiments conducted?	
8	Root causes verified?	
9	Corrective actions tested?	
10	Corrective actions recorded?	
11	DFMEA adjusted?	
12	PFMEA adjusted?	
13	Control plan adjusted?	
14	Work instructions adjusted?	
15	Posted instructions adjusted?	
16	Workforce trained?	

Figure 14.18 Problem-solving acumen audit sheet.

Process capability	Date: _____ Auditor: _____	
Audit daily	**List corrective actions on reverse side!**	
Items to be audited as applicable	**Briefly list discrepancy or condition of concern**	
1	Capability an Issue? (If not an issue then go to next sheet.)	
2	Is there a capability procedure?	
3	All critical dimensions and characteristics have an acceptable Cpk >1.33?	
4	Supplier understands the critical dimensions that require acceptable statistical control?	
5	Cpk calculations are verified with acceptable current analysis?	
6	Is there a plan to improve noncompliant capability?	
7	Is sorting to be used to meet specifications until capability can be achieved?	
8	Is there assurance that sorting will be effective?	
9	Will sorting be done by manual or mechanized means?	
10		
11		
12		
13		
14		
15		
16		

Figure 14.19 Process capability audit sheet.

Process improvements	Date: _____ Auditor : _____	
Audit daily	**List corrective actions on reverse side!**	
Items to be audited as applicable	**Briefly list discrepancy or condition of concern**	
1	Multiple function team evaluates and establishes PFMEA?	
2	PFMEA includes safety issues?	
3	PFMEA includes rework units?	
4	PFMEA includes exception work?	
5	PFMEA improvements are addressed and ongoing?	
6	PFMEA improvements are assigned by impact severity?	
7	PFMEA items are assigned to individuals for correction?	
8	PFMEA items contain a due date for completion?	
9	PFMEA llists temporary actions to overcome temporary difficulties?	
10	All PFMEA revisions are approved by the customer before installation?	
11	There is evidence of ongoing PFMEA improvements?	
12	No other concerns?	
13		
14		
15		
16		

Figure 14.20 Process improvement audit sheet.

Production approval		Date: _____ Auditor: _____	
Audit daily		**List corrective actions on reverse side!**	
IItems to be audited as applicable		**Briefly list discrepancy or condition of concern**	
1	Is there a meeting for design function and production?		
2	Is a design checklist used and completed for product?		
3	Are preproduction process plans developed and approved?		
4	Are DFMEA processes used?		
5	Are PFMEA processes used?		
6	Evidence of corrective actions is shown on current PFMEAs?		
7	Do these functions meet the required supplier standards?		
8	Are lockboxes considered for use to control defective material?		
9	Any other visible concerns?		
10			
11			
12			
13			
14			
15			
16			

Figure 14.21 Production approval audit sheet.

Quality assurance areas		Date: _____ Auditor: _____
Audit daily		**List corrective actions on reverse side!**
Items to be audited as applicable		**Briefly list discrepancy or condition of concern**
1	Inspection area is clean?	
2	All gauges have current calibration?	
3	Work areas are neat and orderly?	
4	If SPC charts are being used, do they have correct control limits?	
5	Are blueprints up to date with correct revisions being used?	
6	Are coordinate measuring machine Programs up to the latest revision?	
7	Are current customer problems displayed and identified?	
8	Are first-piece inspection parts tagged and identified?	
9	Are job instructions posted?	
10	Do the quality assurance staff understand the job instructions?	
11	Are checksheets being used?	
12	Are checksheets complete?	
13	A control procedure for defects?	
14	Are all procedures being followed?	
15	Safety equipment being used?	
16	Are there other visible concerns?	

Figure 14.22 Quality assurance area audit sheet.

Response time		Date: _____
		Auditor: _____
Audit daily		**List corrective actions on reverse side!**
Items to be audited as applicable		**Briefly list discrepancy or condition of concern**
1	Response procedure stated?	
2	Individual assignmments made?	
3	Team activity is mandated with responsible individuals?	
4	List of appropriate actions reviewed with each problem?	
5	At a minimum product is held at: Plant, dock, shipping, trucking, warehouse, etc.?	
6	Problem product contained as per established list?	
7	All customer disruptions and sequesters are scrutinized?	
8	Customer(s) are immediately notified of sequestration?	
9	Customer(s) are updated on status of problem daily?	
10	DFMEA, PFMEA, control plans, work instructions, and applicable postings are corrected?	
11	No other concerns?	
12		
13		
14		
15		
16		

Figure 14.23 Response time audit sheet.

Shipping areas	Date: _____ Auditor: _____
Audit daily	**List corrective actions on reverse side!**
Items to be audited as applicable	**Briefly list discrepancy or condition of concern**

1	Is area clean?	
2	Shipping labels attached properly?	
3	Parts have required part number?	
4	Labels filled in completely?	
5	Containers have approved authorizations?	
6	Parts banded as required?	
7	Parts shrink-wrapped as required?	
8	Scales calibrated?	
9	Foreign material present?	
10	Checksheets being used?	
11	Checksheets maintained?	
12	Procedures being followed?	
13	Safety equipment in use?	
14	Any other concerns?	
15		
16		

Figure 14.24 Shipping area audit sheet.

Tier 2 supplier support		Date: _____ Auditor: _____
Audit daily		**List corrective actions on reverse side!**
Items to be audited as applicable		**Briefly list discrepancy or condition of concern**
1	Supplier directs their suppliers?	
2	This supplier requires Tier 2 suppliers to conform to their requirements?	
3	Supplier aids Tier 2 suppliers in problem solving?	
4	Supplier aids Tier 2 suppliers in developing a continuous improvement system?	
5	Supplier demands that Tier 2 suppliers conform to *no change without preapproval*?	
6	No other concerns?	
7		
8		
9		
10		
11		
12		
13		
14		
15		
16		

Figure 14.25 Tier 2 supplier support audit sheet.

Tool room	Date: _____ Auditor: _____	
Audit daily	**List corrective actions on reverse side!**	
Items to be audited as applicable	**Briefly list discrepancy or condition of concern**	
1	Area is clean?	
2	Temperature controlled?	
3	Humidity controlled?	
4	Calibration results available?	
5	Are calibration records timely and complete?	
6	Job instructions posted?	
7	All personnel certified to perform calibrations?	
8	All gauges clearly identified?	
9	Is there a mold or pattern approval process in effect and being used?	
10	Gauge check-in/check-out procedure being used?	
11	Checksheets being used?	
12	Checksheets complete?	
13	Defect control procedure present?	
14	All procedures being followed?	
15	Any other visible concerns?	
16		

Figure 14.26 Tool room audit sheet.

Unauthorized changes	Date: _____
	Auditor: _____
Audit daily	**List corrective actions on reverse side!**
Items to be audited as applicable	**Briefly list discrepancy or condition of concern**
1	Change procedure exists?	
2	No changes made without customer preapproval?	
3	Customer to approve any plans for proposed changes?	
4	Authorized changes require bank-sized protection?	
5	Failed authorized trial changes have backup protection?	
6	Supplier's supplier(s) aware of unauthorized change rules?	
7	Supplier's supplier(s) aware that inconsequential change approval is required?	
8	No other concerns?	
9		
10		
11		
12		
13		
14		
15		
16		

Figure 14.27 Unauthorized changes audit sheet.

Workstation	Date: _____ Auditor: _____
Audit daily	**List corrective actions on reverse side!**
Items to be audited as applicable	**Briefly list discrepancy or condition of concern**
1	There is adequate lighting?	
2	Workbenches are adjustable?	
3	Fixtures used for assembly?	
4	Jigs used for assembly?	
5	Designated tools stored in designated areas?	
6	Only designated gauges stored in designated spaces?	
7	Job instructions posted?	
8	Photos or sketches of work to be accomplished posted?	
9	No ergonomic problems?	
10	Rubber mat on concrete floor?	
11	Operators can stop process?	
12	Lights or bells used to communicate troubles?	
13	Environment-friendly?	
14	No other concerns?	
15		
16		

Figure 14.28 Workstation audit sheet.

chapter fifteen

Summary

In summary, there are many easy steps that can be employed to facilitate supplier enhancement. Suppliers should be considered to be valued partners because they have a significant impact on the profitability of both their and the customers' top and bottom lines. Unfortunately, some suppliers do not have the experience necessary to provide a qualified product economically. If a supplier is not sufficiently capable or sophisticated, the costs associated with the product they provide to the customer can be excessive. In other cases, the provision of substandard services or products can result in final customer dissatisfaction that will lead to reduced sales. These expensive and unnecessary costs can be reduced through the application of supplier mentoring. The effort involved in advising suppliers will have a direct impact on the performance and profitability of the customer and supplier organizations.

There are many different methods of enhancing the supplier base. One of these is the use of mentoring when it is required. The style of supplier tutoring is dependent on whether the supplier is internal or external to the customer organization. Internal suppliers—that is, operations that precede a following operation—can be corrected within the supplier base by providing employee training. This training involves the application of the *Trilogy of Supplier Enhancement* philosophy. There is a further requirement to apply the *Why Sequence* as an economical means of reducing scrap, rework, and unnecessary work. In addition to determining why a problem was caused, these three problem-solving steps require answers as to why the problem was not captured and why the problem was not prevented in the planning stages. External suppliers, on the other hand, must be guided to accept their responsibility to provide the customer with a quality low-cost product or service. There are many means to guiding external suppliers, including providing up-front expectations and holding quality meetings or symposiums that suppliers are required to attend. All of these facets of requiring supplier services must be recognized and addressed. These include

1. A supplier agreement that absolutely no changes will be tolerated once a contract is established without prior customer approval. This includes any changes to secondary suppliers, materials, tools used, location, methods, controls, tests, operation sequences, heat treatment, or any other anomaly not previously approved.

2. The incorporation of the Trilogy of Supplier Enhancement philosophy into mandated use and application by the entire workforce; the use of the Why Sequence to solve problems.
3. The practice of keeping accurate records that facilitate problem-solving skills by not only revealing why the current practice has failed but also why it was not recognized and why it was not prevented.
4. The creation and application of supplier audits to ascertain that new and experienced suppliers are qualified, competent, and effective in their actions; that they possess the credentials, elements, record-keeping, readiness, and problem-solving skills to effectively service the customer.
5. The use of audits to justify the continued acceptance of the methods employed by the supplier and the customer.
6. Recognition that it is necessary to trust suppliers but it is also necessary to verify their actions.

Now is the time to begin or to continue supplier enhancement. It is not difficult or costly to apply these concepts. The understanding and application of the above six precepts will greatly improve the effectiveness of a customer organization. Of course, there is a necessity to ensure that all corrective actions are maintained. Accurate record keeping and the use of a champion quality assurance philosophy will ensure success and provide an acceptable means of creating a *lessons learned* attitude that can be applied in other areas.

So, in review, the following information has been illustrated.

1. Supplier differences between internal and external suppliers are discussed and the common methods to improve both their performance and relationships are provided.
2. Techniques are given to simplify supplier problems and to innovate significant solutions.
3. The different methods employed by qualified suppliers compared with those less experienced are espoused.
4. The materials presented provide an understanding of how supplier improvements can substantially improve top- and bottom-line performance.
5. Examples, photos, forms, lists, and audit worksheets provide information and illustrate how to take a quality management approach to facilitate supplier improvements.
6. A series of worksheets is provided that can be used to establish an audit enhancement base at the customer or supplier organization.

I sincerely hope that you find this information to be useful. May you have great success in its application.

Appendix

The following terms were used in the samples and explanations provided.

Audit A visual observation of the conditions present to determine if items requiring compliance are being followed.

CFR The document and requirements to comply with the Code of Federal Regulations.

Control plan The designated system to be employed to manufacture a product or provide a service in a specified manner, which includes the methods, checks, tests, and criteria to be used.

Corrective actions Those actions that are to be taken to minimize or eliminate adverse conditions from occurring again once a problem is recognized.

Dead time A term used to denote the period when an operation cannot be performed due to assignable but preventable causes.

Defect Any anomaly that detracts from the acceptability of a product or service. It might be a flaw, a feature, or an occurrence.

DFMEA Design failure mode and effect analysis. An instrument used in the design phase of a product that considers the failure modes associated with the design of the product.

DPMO Defects per million opportunities. An indication of the likelihood of defects occurring.

External supplier Those suppliers that are from outside of the customer organization.

Internal supplier Those suppliers that are from inside the customer organization and are located upstream of the manufacturing or service process.

ISO The International Organization for Standardization, which develops and publishes international standards.

Lockbox A device that is used to capture and hold a defective part until an evaluative determination can be made.

No change without prior approval The understanding that a product, service, or method, no matter how insignificant, will not be made without the authorization of the customer base.

Perusal A cursory inspection or glance at a component or service to determine its acceptability or obvious deficiencies.

PFMEA Process failure mode and effect analysis. An instrument used in the process planning phase of a product or service that considers the failure modes associated with and the methods used to control the outcome of the product used.

PPM Parts per million, as a frequency of occurrence.

Problem data sheet A problem-solving tool that is developed by the customer and the supplier to supply all the pertinent information available to define the current problem under review.

Problem resolution sheet A form that can be used to gather important information relative to the problem at hand.

Process parameters Those data or other information that describe the limits and operation of the process.

QS Quality system, established and developed to control the quality output of a service or product.

Sequester The impounding of suspect product that requires relocation to a secure area to prevent it from being processed or shipped until a determination can be made.

Spill A condition that allows defective materials to be shipped to a customer.

Supplier Those involved in supplying a component, product, or service that is upstream of the current operation.

Symposium A customer-provided supplier meeting where differences are resolved relating to supplier quality performance.

Table of comparison An analysis tool that can be used to differentiate components when a comparison is rendered between samples of five, allowing a decision to be made.

Tier 2 supplier A supplier that provides a product to a customers' main supplier—that is, a supplier two stations upstream.

Trilogy of Supplier Enhancement A trio of requirements that can lead to the improvement of internal and external suppliers. These include a cursory review of supplied parts, the requirement to make good product, and the sequestration of questionable inputs, products, or services.

Why Sequence A structure that requires not only the cause of the problem to be determined but also requires a determination of why the problem was not prevented in the planning process and why the defect was not captured when it was created.

Work instructions Individual workstation or employee instructions that designate how the work is to be performed with the tools and methods specified.

Worksheets Sample audits that can be used as a basis for creating and conducting applicable audits.

Index